Dinosaurium

For Jack, Daisy and Eliza – C.W.
For O & A and all chasers of dinosaurs – L.M.

BIG PICTURE PRESS

First published in the UK in 2017 by Big Picture Press,
an imprint of Kings Road Publishing,
part of the Bonnier Publishing Group,
The Plaza, 535 King's Road, London, SW10 0SZ
www.bigpicturepress.net
www.bonnierpublishing.com

1 3 5 7 9 10 8 6 4 2

ISBN 978-1-78370-792-8

This book was typeset in Gill Sans and Mrs Green.
The illustrations in this book are digital engravings.

Designed by Winsome d'Abreu
Edited by Tasha Percy

Printed in China

This book was produced in consultation with Dr Jonathan Tennant.

Welcome to the Museum
ENTER HERE

Dinosaurium

Illustrated by CHRIS WORMELL

Written by LILY MURRAY

BPP

Foreword

Dinosaur fossils have been discovered all around the world across many different habitats. With each new discovery new questions are raised and our understanding of them develops.

The pace of dinosaur discovery is getting faster year on year. Although the first dinosaurs were named nearly two centuries ago in 1824, more than half of all known dinosaurs have been named in the last three decades. Today we know more about dinosaurs than ever before: there are more museums and palaeontologists than at any point in history, and new discoveries are being made all the time. The study of dinosaurs is no longer solely focused on fossil bones and reconstructing skeletons. Research now includes exploring evidence of protein molecules, bone and eggshell microstructure, feather and skin traces, footprints and trackways as well as x-rayed limbs and skeletons.

Journey back through prehistory to witness the break-up of the supercontinent Pangaea, and the coming of the dinosaurs. Your tour of their world begins with a dinosaur family tree, which shows how these remarkable creatures evolved to dominate the Earth for 175 million years. The detailed illustrations that follow bring these remarkable creatures to life in all their glory — savage hunters with razor-sharp claws, lumbering plant-eating giants and feathered ancestors of today's birds.

We are experiencing a renaissance of dinosaur discovery and research, which has outdated some of the earlier science and understanding of these prehistoric beasts. So use your journey through the dinosaur world in the pages that follow as food for thought to build your interest. How much can we really know about dinosaurs? Which dinosaur was the longest? Could they swim and fly? Who had the longest claws? How are dinosaurs and modern birds connected? All this and more will be revealed within.

Professor Paul Sereno
Palaeontologist, University of Chicago

Entrance

Welcome to Dinosaurium

This museum shows you life on Earth as you've never seen it before. It will transport you millions of years back in time to discover the largest and most ferocious land animals that ever lived — the dinosaurs. Marvel at the museum's extraordinary catalogue, which brings you these creatures in all their wondrous variety, from tiny feathered killers to vast plant eaters that shook the earth as they walked.

As you wander through the pages of this book, you will tour galleries that reveal how dinosaurs lived and how they changed over time. Discover the astounding variety of dinosaur species and find out what they ate, how they moved, where they lived and how they fought. See for yourself how dinosaurs evolved from their most primitive forms to a vast array of species and read the amazing story of how dinosaurs evolved into birds.

Look carefully at each exhibit. Some have dioramas that recreate the world as it was in dinosaur times, when strange plants covered the Earth and the largest mammal was the size of a shrew. Others will show you dinosaur skeletons, amazing fossil finds and how the continents have moved over time.

Passing through the galleries, find out how different groups of dinosaurs are related to each other and how feathers evolved. Be inspired to imagine the world millions of years ago, when reptiles ruled the Earth.

Enter Dinosaurium and discover these incredible, terrifying, ingenious creatures, brought back from the mists of time. Welcome. This museum is yours to treasure.

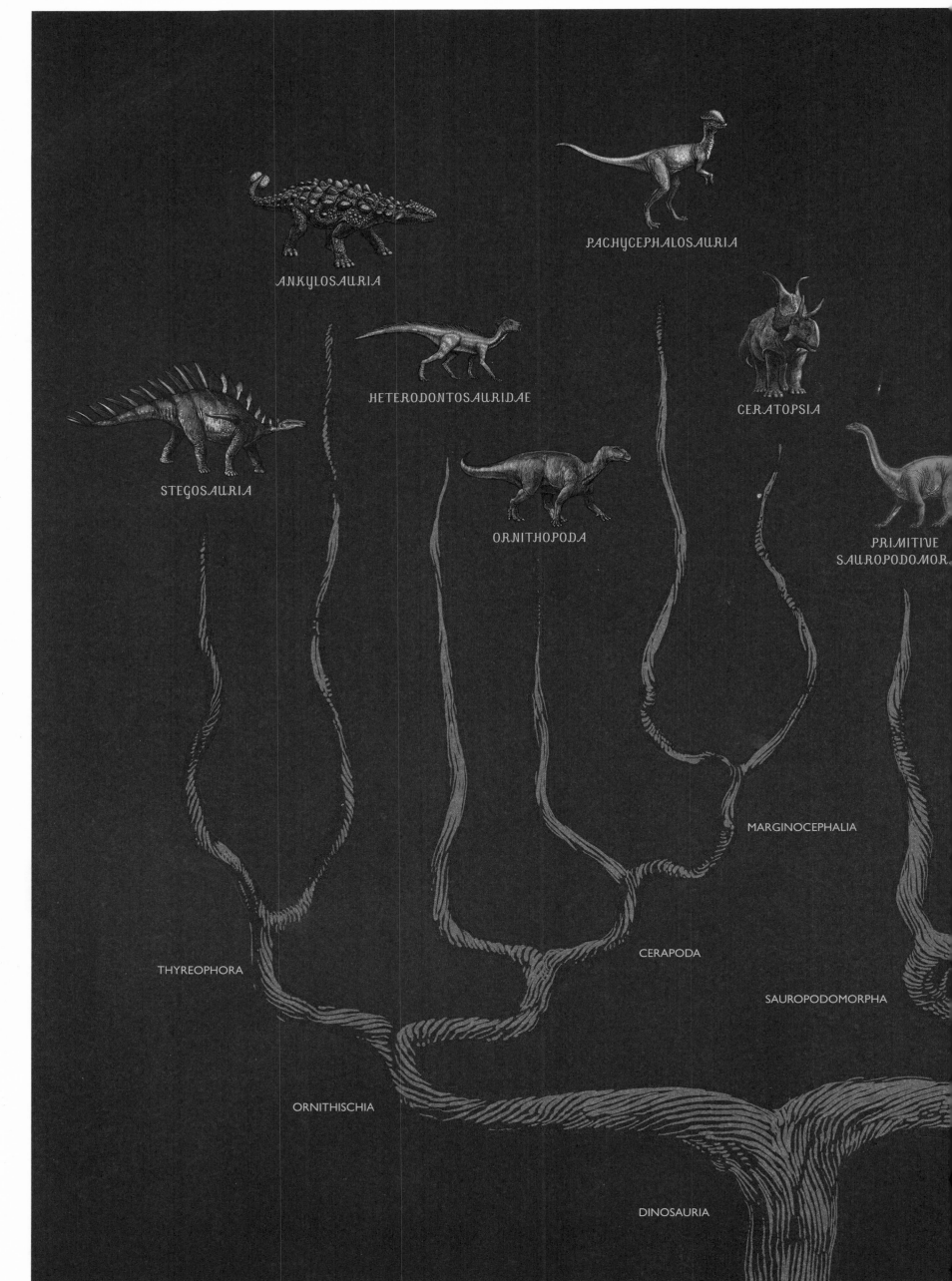

ANKYLOSAURIA

PACHYCEPHALOSAURIA

HETERODONTOSAURIDAE

CERATOPSIA

STEGOSAURIA

ORNITHOPODA

PRIMITIVE
SAUROPODOMOR

MARGINOCEPHALIA

THYREOPHORA

CERAPODA

SAUROPODOMORPHA

ORNITHISCHIA

DINOSAURIA

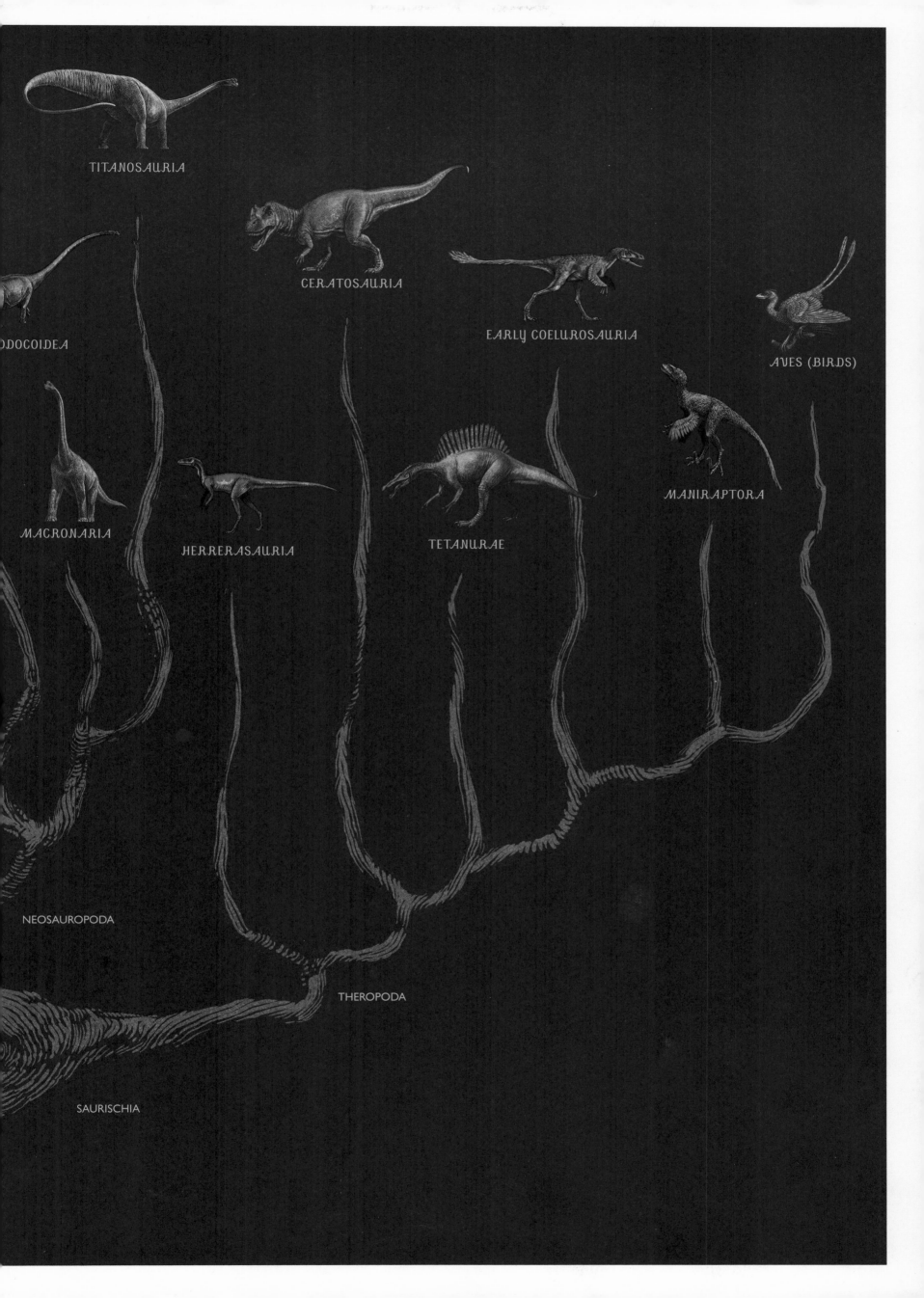

TITANOSAURIA

DOCOIDEA

CERATOSAURIA

EARLY COELUROSAURIA

AVES (BIRDS)

MACRONARIA

HERRERASAURIA

TETANURAE

MANIRAPTORA

NEOSAUROPODA

THEROPODA

SAURISCHIA

Dinosaur classification

The dinosaur family tree, known as a cladogram, shows how groups of dinosaurs were related to each other. It also visualises the huge variety of forms that evolved, from the first scaly two-legged dinosaurs to huge four-legged beasts and elegant flyers.

Dinosaurs are part of a group of reptiles called archosaurs that evolved over 250 million years ago. They were first classified as a distinct group in 1842 by palaeontologist Sir Richard Owen, when he placed *Iguanodon*, *Megalosaurus* and *Hylaeosaurus* in "a distinct tribe or suborder of Saurian Reptiles, for which I would propose the name of Dinosauria."

In 1887 and 1888, dinosaurs were further divided into the saurischians and the ornithischians, with the distinction based on the structure of their hip bones. The saurischians had hip bones similar in shape to those of modern lizards, with a pubis bone (one part of the three bones that make up the hip bone) that pointed forward, while the ornithischians had hip bones similar to those of modern birds, with a pubis bone that pointed backwards. Since then, over 900 dinosaur species have been discovered, with new species being discovered almost every couple of weeks, but the basic structure of dinosaur classification remains the same. Recently, however, this classification has been challenged, and theropods and ornithischians might be more closely related to each other than previously thought.

Dinosaurs are further categorised into smaller groups, or clades, consisting of an ancestor and all of its descendants. Each clade is made up of dinosaurs that share certain features, such as the similar wrist joints of maniraptorans, or the distinct frills of the ceratopsians. This system of classification helps scientists to study the evolutionary relationships between different groups of dinosaurs and led to the ground-breaking discovery that dinosaurs are not extinct, as was once thought, but alive and flourishing today as birds.

The Mesozoic Era

The Mesozoic era, also known as the 'Age of Reptiles', lasted from around 252 to 66 million years ago. It is divided into three periods: the Triassic, the Jurassic and the Cretaceous.

Dinosaurs first appeared in the Triassic period, around 240 million years ago, when the Earth looked very different to how it does today. At the beginning of the Triassic period, most of the continents were joined together as one huge supercontinent called Pangaea, which covered around one-quarter of the Earth's surface. Surrounding Pangaea was the vast Panthalassa Ocean, while the smaller Tethys Ocean wrapped itself around Pangaea's eastern coastline. By the end of the Triassic period, some parts of Africa, North America and Europe had begun to drift away from each other, and the North Atlantic Ocean began to form.

During the Jurassic period, Pangaea split into two major subcontinents, to form Laurasia in the north and Gondwana in the south. By the Middle Jurassic, Gondwana was beginning to break up, the eastern part (Antarctica, Madagascar, India and Australia) splitting from the west (Africa and South America). In the north, the North Atlantic Ocean also continued to widen and North America and Eurasia drifted further apart. Mountains rose on the sea floor, pushing sea levels higher on the continents, which in turn produced a more humid, wetter climate.

For much of the Cretaceous period, high sea levels meant that large parts of the continents were underwater, although there was occasionally a land link between North America and Asia, across which dinosaurs would have migrated. The continents continued to drift apart so that by the Late Cretaceous, most of the major continents were separated by oceans and looked much like the continents do today.

Key to plate

1: Triassic period

The Triassic period lasted from 252 to 201 million years ago. At the beginning of the Triassic, only China and parts of southeast Asia were separate from Pangaea. This image of the Earth, in the Late Triassic, shows the first signs of rifting in Pangaea and the emergence of the North Atlantic Ocean. Whereas coastal areas were green, large parts of the interior were covered in desert.

2: Jurassic period

Following on from the Triassic period, the Jurassic lasted from 201 to 145 million years ago. On this map you can see the beginning of the break-up of Pangaea and how rising sea levels flooded large parts of the continents. It was also at this time that mountains such as the Rockies of North America, the Andes of South America and the Alps of Europe were formed.

3: Cretaceous period

The last and longest of the Mesozoic periods, the Cretaceous period lasted from 145 to 66 million years ago and saw the final break up of the continents. This reconstruction of the Earth from the Late Cretaceous period also shows the Western Interior Seaway, which divided North America, and the large inland sea covering North Africa.

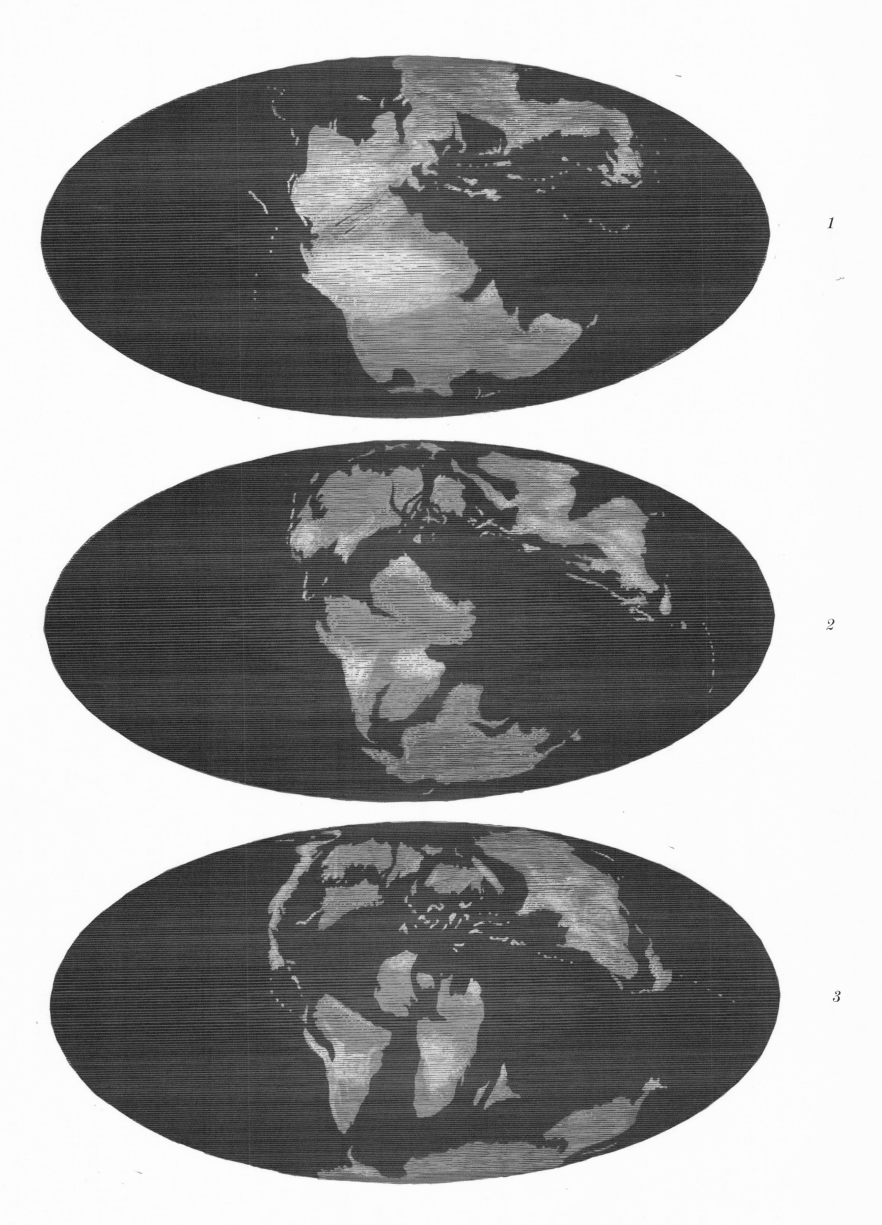

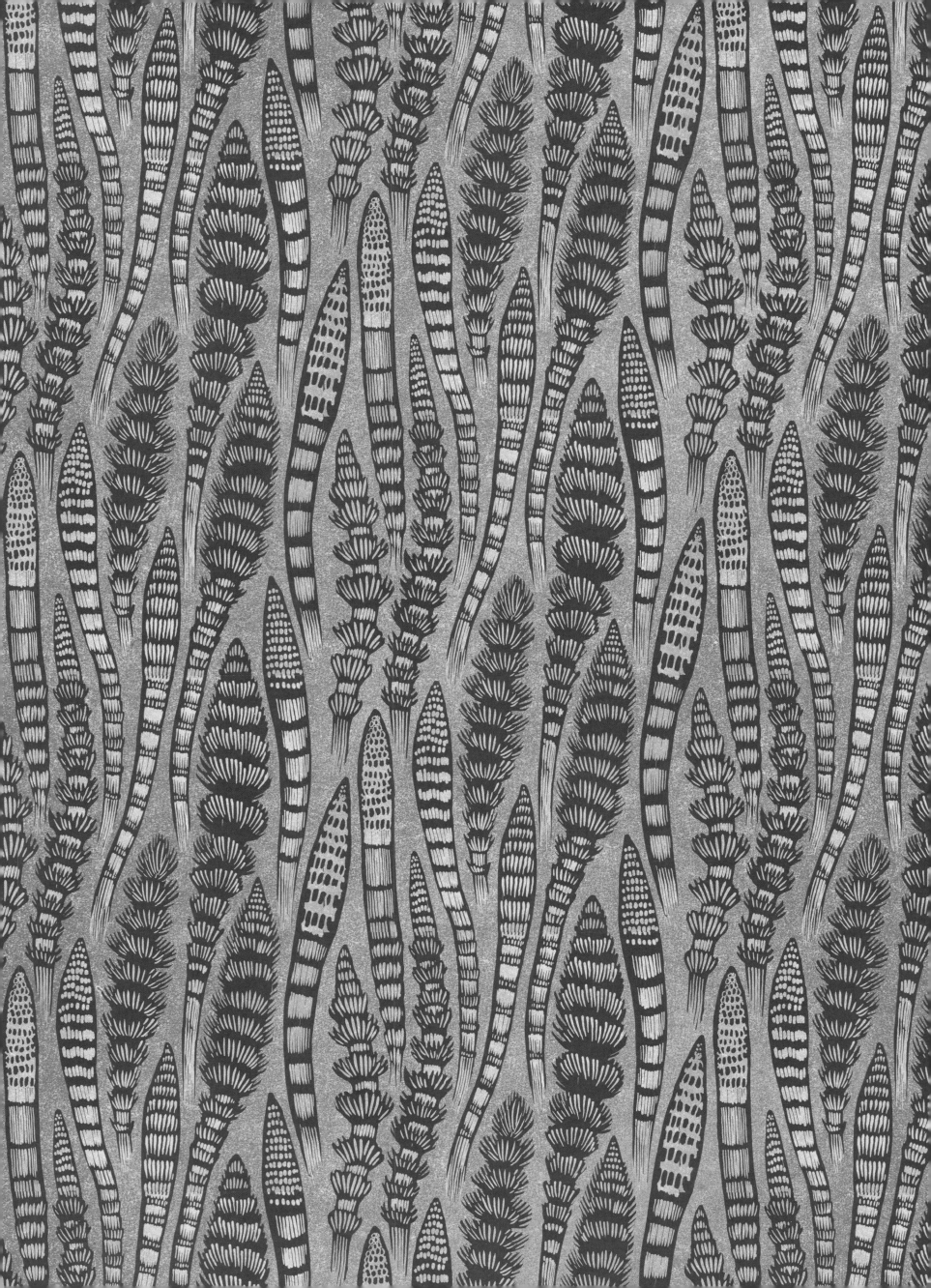

Sauropodomorpha

Sauropodomorpha

Very early in the evolution of the dinosaurs, a branch of saurischian dinosaurs, known as the sauropodomorphs, broke away from the predatory theropods and became plant eaters. This branch, which includes the sauropods and their ancestral relatives, lasted from the Late Triassic until the very end of the time of the dinosaurs.

To begin with, the sauropodomorphs were relatively small and walked on two legs. These early forms were most likely omnivorous. They had massive, clawed thumbs, probably used for defence as well as to pull down branches for feeding. Over time, however, they grew larger, and their increasing size forced them to spread their weight across four pillar-like legs. By 220 million years ago, the sauropods had become Earth's dominant large terrestrial herbivores.

The sauropodomorphs are characterised by their long necks, which they evolved in order to browse, giraffe-like, high in the treetops, accessing foliage other herbivores could not reach. In this they were aided by small, light skulls and a long counterbalancing tail.

Their leaf-shaped teeth could easily slice through tough stems but were unsuited to grinding up food. Instead, they had stomach stones, known as gastroliths, similar to the gizzard stones of modern birds, to help digest tough plant fibres. There is also evidence that the mouths of some species may have ended in a small beak.

Their fossils have been found across all continents and a range of environments, from swamps to deserts. The largest sauropodomorphs have come to symbolise the meaning of the word 'dinosaur' in the popular imagination – unimaginably vast animals that were taller than buildings, longer than buses and whose footsteps literally shook the ground as they walked.

--- *Key to plate* ---

1: **Brachiosaurus altithorax**
Late Jurassic, North America
Length: 25m; Weight: 28,000kg
When its fossils were discovered in 1900, *Brachiosaurus* took the record of the largest dinosaur. Although it has now been surpassed, it is still one of the tallest known dinosaurs.

Brachiosaurus had a giraffe-like body, with a long neck and unusually long front limbs that may have sprawled

outwards. Its name actually means 'arm lizard'. This giant of a dinosaur could have been even larger than we think, as the lack of fusing in some *Brachiosaurus* bones suggest that the specimen wasn't even fully grown.

Brachiosaurus would have used its long neck to reach leaves, as it would have been unable to rear up on its hindlimbs. In order to sustain itself, it would have needed to eat up to

120kg of cycads (an ancient group of seed plants), conifer and ginkgo leaves every day.

2: **Brachiosaurus altithorax skull**
The dinosaur's skull had a wide muzzle and thick jawbones that housed 52 spoon-shaped teeth (26 in each jaw), which were perfectly suited for stripping vegetation. The large cavity at the top of its head housed its nostrils.

Primitive sauropodomorpha

Once grouped together as the ancestors of the giant sauropods, these dinosaurs are now thought to be early relatives. They date from the very dawn of the dinosaurs – their fossils are some of the world's oldest discovered dinosaur bones, dating back to 200 to 225 million years.

From the Late Triassic to the Early Jurassic, these were the most common herbivores of their day, and the first group of dinosaurs to dominate their environment. Their fossils have been found all over the world, even in Antarctica, although most are known from northern Europe. They would have browsed for food among the high branches, rearing up on their back legs to reach the best foliage. However, by the mid-Jurassic they had disappeared from the fossil record, possibly dying out as their relatives, the sauropods, outcompeted them for food.

The evolution of the sauropodomorphs shows how they progressed towards larger body sizes, smaller heads, longer necks, and walking on four legs. This would have helped them reach higher into trees to feed and provided defence against the increasing size of their predators – the theropods.

——————————— *Key to plate* ———————————

1: Massospondylus carinatus
Early Jurassic, Zimbabwe and USA
Length: 4m; Weight: 135kg
This species had a longer neck than most other primitive sauropodomorphs. It also had massive thumb claws which it may have used to tear off branches or roots. *Massospondylus* eggs show that hatchlings had no teeth and were clumsy walkers, suggesting adults would have cared for their young.

2: Plateosaurus engelhardti
Late Triassic, Germany, Switzerland and France
Length: 10m; Weight: 4000kg
One of the best known European dinosaurs. Hundreds of fossils have been found together in one place, suggesting they lived in herds.

3: Thecodontosaurus antiquus
Late Triassic, England
Length: 2.5m; Weight: 40kg

Far smaller than its near contemporary, *Plateosaurus*, it is now thought to be a 'dwarf' island species. It was the fourth dinosaur to be named.

4: Riojasaurus incertus
Late Triassic, Argentina
Length: 6.6m; Weight: 800kg
With its large body and bulky legs, *Riojasaurus* was a slow-moving animal that was probably unable to rear up on its back legs.

4

The Triassic Period

Around 251 million years ago, there was a mass extinction in which an incredible 96 per cent of all life forms died out. The Triassic period that followed saw a major growth of life on land, with both the early ancestors of mammals and dinosaurs appearing for the first time.

At the beginning of the Triassic, temperatures were warmer than they are today. There was no ice on the poles and a vast desert covered the interior of Pangaea. On higher, cooler ground, gymnosperms (plants with exposed seeds) could be found as well as coniferous forests.

The climate around the coast was now much wetter, and it was here that most life existed. There were mosses and ferns, spiders, scorpions, millipedes, centipedes and beetles. The Triassic also saw the appearance of the first grasshoppers.

The largest life forms on land were mammal-like reptiles, known as therapsids, and the archosaurs. By the mid-Triassic, the archosaurs had branched into the first dinosaurs, and by the Late Triassic, the winged pterosaurs, the first vertebrates capable of active flight.

The earliest mammal ancestors evolved at the very end of the Triassic, from the therapsids. These were tiny, shrew-like creatures that fed either on plants or insects.

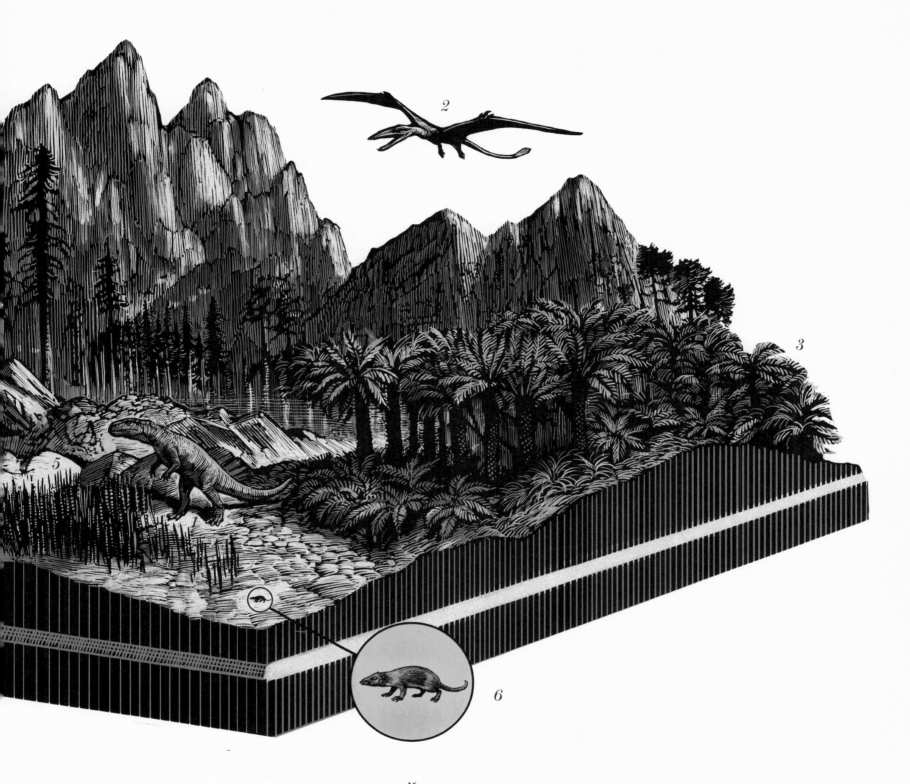

1: Postosuchus

Length: 5m; Weight: 680kg
A top predator in North America, *Postosuchus* was an archosaur with pillar-like upright legs, making it a fast, agile hunter. It lived alongside the small dinosaurs of its time, like *Coelophysis*. Its forelimbs were much shorter than its hindlimbs, suggesting it may have walked on two legs.

2: 'Fanged pterosaur'

Wingspan: 1.3m; Weight: uncertain
Triassic pterosaurs were still relatively small. This one, discovered in 2015 and as yet unnamed, had 110 teeth and four 2.5cm-long fangs. It would

have been capable of short flights and preyed on insects and tiny ancient ancestors of crocodiles.

3: **Bennettitales**

These palm-like plants flourished during the Triassic. They had tough leaves and woody trunks, with short, barrel-shaped stems.

4: Araucarioxylon arizonicum

A species of conifer, forests of which covered North America in the Late Triassic. Its closest relative today is the monkey puzzle tree.

5: **Horsetails**

These rush-like plants were an important food source for the herbivores of the time. They evolved new forms during the Triassic. They reproduced by spores rather than seeds, and were fast-growing and resilient, with underground stems.

6: Morganucodon

Length: 13cm; Weight: 27–89g
An early mammal ancestor, *Morganucodon* still had reptilian features, including the shape of its jaw. It laid eggs, which were probably small and leathery, and was most likely nocturnal.

Sauropoda

A hugely successful herbivorous group, the first true sauropods evolved at the end of the Triassic and survived for more than 100 million years until the end of the Cretaceous. They include the largest land animals ever to have lived and at their heaviest weighed close to 100,000kg – 16 times the weight of an African elephant. Often various species went on to develop extremely large bodies. But *how* did they get so big?

A key feature was their long necks, which enabled them to outcompete other herbivores by accessing a wide range of food sources. The length of the neck was made possible by their light, hollow bones, filled with air, which reduced the overall weight. For support, they had up to 19 neck vertebrae and specially arranged muscles, ligaments and tendons. They could eat quickly too, swallowing their food without chewing, and their small heads were really all-mouth, making the sauropods a lot like Mesozoic lawnmowers.

It is likely they also had a highly efficient breathing system, like that of modern birds,

with a system of air sacs that meant they could pump air through their lungs in a continuous flow, instead of having to breathe out and then in again like mammals do.

Key to plate

1: Mamenchisaurus hochuanensis
Late Jurassic, China
Length: 25m; Weight: 36,000kg
Famous for its extraordinarily long neck, which took up half its entire body length. Some scientists argue it would have had a near horizontal neck posture and browsed at low or medium heights.

2: Diplodocus carnegii
Late Jurassic, USA
Length: 22–35m; Weight: 18,000kg
One of the longest known dinosaurs, in large part because of its whip-like tail, which it may have used to defend itself. A *diplodocus* could swing its tail fast enough to create a loud cracking noise to warn off predators.

3: Nigersaurus taqueti
Mid-Cretaceous, Niger
Length: 9m; Weight: 2000kg
One of the more unusual sauropods with incredibly wide jaws holding around 600 needle-shaped teeth. It probably fed by swinging its neck from side to side, clipping its way through low foliage as it went.

4: Amargasaurus cazaui
Early Cretaceous, Argentina
Length: 13m; Weight: 4000kg
Instantly recognisable by its row of forked spines, which may have been for temperature regulation or defence.

Titanosauria

The beginning of the Cretaceous period, 145 million years ago, saw the decline of some of the sauropods, such as the Diplodocidae and Brachiosauridae. But this did not mean that the sauropods as a whole were heading for extinction. The time also marked the rise of a new branch of the sauropods – the titanosaurs, which included some of the largest land-living animals that ever existed. Their fossils have been found all across the world, although the majority have come from South America, which was then part of the giant landmass Gondwana. A titanosaur discovery in 2014, from Patagonia, is among the largest dinosaurs discovered so far.

Titanosaurs stood on four column-like legs and walked on toes that had fused into horseshoe-shaped stumps of bone, with huge fleshy pads on their feet to cushion the impact as they moved. Like the sauropods, the titanosaurs were long-necked and long-tailed herbivores that used their teeth to strip leaves from trees, but could not chew.

Interestingly, many titanosaur species were armoured with small, bead-like scales along their backs, although one species, *Saltasaurus*, even had bony plates, like the ankylosaurs. These could have been for defence against the giant theropods, such as the abelisaurids and the tyrannosaurs, the super-predators of the day.

───────────────── *Key to plate* ─────────────────

1: **'The Titanosaur' (not yet named)**
Late Cretaceous, Argentina
Length: 37m; Weight: 70,000kg
This species, as yet unnamed, is known from seven skeletons found in 2014, in Patagonia. Currently referred to as 'the Titanosaur', it weighed as much as 10 African elephants and was longer than three London buses end to end. Scientists have estimated it would have been 10 per cent larger than *Argentinosaurus*, the previous record

holder. Seven of the animals were found together, which means they must have died at exactly the same spot. It also seems as if the titanosaurs may have been scavenged after death. Over 50 teeth from *Tyrannotitan chubutensis*, one of the largest known theropods, were also discovered at the fossil site. These lucky theropods may have been able to feast for weeks on the titanosaurs' remains.

When walking, the creature's long

neck would have been held horizontal. But when feeding, its neck could have reached upwards to browse at a height of up to 14m, or swooped down low to feed on plants close to the ground. In order to survive, this animal would have had to eat massive quantities of plants a day, which it would then digest in its enormous gut.

Theropoda

Theropoda

Theropods, meaning 'beast-footed', were a diverse range of largely flesh-eating dinosaurs that walked on two legs. Like the sauropodomorphs, the theropods were a branch of saurischian ('lizard-hipped') dinosaurs, and first appeared in the Late Triassic period, around 230 million years ago. The dominant land predators of their time, they ranged from the crow-sized *Microraptor* to the ferocious, earth-pounding giants like *Tyrannosaurus* and *Giganotosaurus*. They also included omnivores, insectivores and even herbivores, such as the strange, razor-sharp clawed therizinosaurs.

Technically, theropods still survive today, as all modern birds are descended from small, non-flying theropod dinosaurs. In fact, from the Jurassic period, many theropods were beginning to look increasingly like birds, with beaks, wishbones, and arm and leg feathers that formed wings. We now know that many displayed bird-like behaviour too — many theropods built nests and brooded over their eggs.

Other common theropod features include hollow bones, sharp claws for tearing and grasping prey, and sharp, curved and serrated teeth for cutting through flesh. Like birds, they had four toes but only walked on three. Many species had limited use of their forelimbs, as they could not rotate them, so their palms always faced inwards or towards their legs. Theropods were also often unable to rotate their wrists, so the forelimb and hand had to move as one.

By the Jurassic period, theropods had become the top land predators and remained so for 100 million years, right until the end of the Cretaceous period. Theropod fossils have now been found on every continent. Antarctica was the last continent to yield up theropod remains, with the discovery of *Cryolophosaurus*, a medium-sized theropod with a forward-facing crest, in 1991.

Key to plate

1: Coelophysis bauri
Late Triassic, North America
Length: 3m; Weight: 25kg
This dinosaur, first discovered in 1881, is the best-known theropod of the Late Triassic. In 1947, over 1000 specimens were found together at Ghost Ranch in New Mexico.

It has many key theropod features, including hollow bones (its name means 'hollow form'), grasping hands and sharp teeth and claws. It is also the earliest known dinosaur to possess a wishbone.

Coelophysis was small, but swift and agile, and we know from contents found in the fossilised remains of its stomach that it fed on small crocodilians.

The discovery of so many fossils together suggests that *Coelophysis* may have hunted in packs, but it is just as possible that many had gathered together at a waterhole and then drowed in a flash flood.

2: Coelophysis bauri skeleton
This is a depiction of a fossilised find from the Petrified Forest of Arizona. You can clearly see the three weight-bearing toes on its feet, the smaller fourth toe and the long S-shaped neck (here bent backwards in death).

1

Ceratosauria

The Ceratosauria, meaning 'horned lizards', were a primitive group of theropod dinosaurs, dating from the Late Triassic period. They are loosely defined as a group of theropods more closely related to *Ceratosaurus* than to birds. The Ceratosauria first appeared in the Late Triassic period, around 225 million years ago. One of the best known is *Coelophysis*, a swift, agile hunter, with strong hind legs, a long tail and an S-shaped neck. Many fossil specimens have been found together, suggesting they lived, and possibly hunted, in packs.

Early Jurassic ceratosaurians included the twin-crested *Dilophosaurus* (made famous by the film *Jurassic Park*), as well as *Ceratosaurus* (illustrated above) from the Late Jurassic.

By the Early Cretaceous period the Ceratosauria had disappeared from the northern continents and had spread to Gondwana, in the south. Here they evolved into relatively large carnivores, such as *Abelisaurus*, *Carnotaurus* and *Majungasaurus*. Like *Ceratosaurus*, *Carnotaurus* had horns above its eyes, while the slightly smaller *Majungasaurus* was the apex (top) predator of its time in Madagascar, which was already an island. *Majungasaurus*

would have preyed mainly on sauropods, such as *Rapetosaurus*, and is also one of the only dinosaurs for which direct evidence of cannibalism is known.

Many of the species took on strange forms, including the small, toothless herbivore *Limusaurus,* and *Masiakasaurus* which had teeth that stuck forwards out of its mouth, rather than downwards, most likely in order to catch fish and small vertebrates.

--------------------- *Key to plate* ---------------------

1: Ceratosaurus nasicornis

Late Jurassic, USA
Length: 7m; Weight: 700kg
A medium-sized theropod distinguished by a large horn over its nose and two horn-like ridges above its eyes. It also had a row of bony bumps, known as osteoderms, along its back. Some scientists think the horn could have been used by males to fight over females in breeding contests, while others argue it was too fragile and would have been used for display purposes only, in which case it may have been brightly coloured.

Ceratosaurus lived alongside other, larger theropods, such as *Allosaurus* and *Torvosaurus*. While they probably preyed on sauropods in the region, it's likely *Ceratosaurus* fed on smaller prey, such as ornithopods. It had a light skull with large, blade-like teeth, perfect for slicing through flesh. One specimen was found with teeth so long they extended below the lower jaw when the creature closed its mouth. Its tooth marks have also been found on sauropod bones, showing it could hunt larger animals, or at the very least was prepared to scavenge them.

In the Late Jurassic, southwestern USA was covered in swampland, and it has been proposed that *Ceratosaurus'* long, flexible tail may have been used for swimming, enabling it to hunt in the water for fish and crocodilians.

1

Allosauroidea

This group of fearsome predators is made up of two families – the Allosauridae and the Carcharodontosauridae. The Allosauridae were medium to large carnivores and the most successful hunters of the Late Jurassic. Their most famous member, *Allosaurus*, ruled the American Midwest, with its knife-like teeth, terrifying claws and long, strong legs. Its prey would have included stegosaurs, ornithopods and even sauropods.

The Allosauridae had large, crested skulls and three-fingered hands, and have been found in North America, Africa and Asia. Unlike the Tyrannosauridae, many allosaurids had good-sized, powerful forelimbs for grasping prey. They were eventually succeeded by the Carcharodontosauridae, which along with the spinosaurids, were the dominant predators of Gondwana during the Early and Late Cretaceous, with species also known from North America and Asia.

The Carcharodontosauridae include *Neovenator*, a formidable hunter from the Isle of Wight, England, and some of the largest land predators ever: *Giganotosaurus*, *Mapusaurus* and *Carcharodontosaurus*, which were equal in size or even larger than *Tyrannosaurus*. *Giganotosaurus*, the largest of them all, had a skull the length of a man, which was full of knife-sized, flesh-slicing teeth and a body as long as a bus. It is likely all three massive carcharodontosaurids would have hunted the huge Cretaceous titanosaurs of the southern hemisphere.

The dinosaur *Carcharodontosaurus*, for which the Carcharodontosauridae is named, was very nearly lost to science. The first fossils were discovered by Ernst Stromer in Egypt in

the 1920s and stored in Munich, Germany. They were subsequently destroyed, along with the first remains of *Spinosaurus*, in a bombing raid in World War II, and for many years *Carcharodontosaurus* was only known by Stromer's drawings and descriptions. Then, in 1995, a huge skull and partial skeleton was found in the Sahara Desert, in southeast Morocco, by Paul Sereno, that matched Stromer's description. A year later, Sereno and palaeontologist Steve Brusatte found a second species, *Carcharodontosaurus liguidensis*.

Key to plate

1: Carcharodontosaurus saharicus **skull**
Mid-Cretaceous, Africa
Length: 13m; Weight: 6000kg
Named after *Carcharodon*, the scientific name for the Great White Shark, *Carcharodontosaurus* means 'sharp-toothed' or 'jagged-toothed lizard' and its skull shows why – it was full of enormous serrated teeth, 20cm long. Scans of the skull have shown that the inside inner ear, as well as its brain size, were more similar to modern reptiles than to birds.

Despite being one of the largest land predators of all time, *Carcharodontosaurus* would not have been without rivals. It lived at the same time and in the same region as the 18m-long *Spinosaurus*, one of the mightiest beasts of the Mesozoic era.

2: Allosaurus fragilis
Late Jurassic, USA
Length: 8.5m; Weight: 1700kg
A typical large theropod with a muscular S-shaped neck, large skull, long tail, and forelimbs that were short in comparison to the hindlimbs. All of its teeth had saw-like edges, to help it cut through flesh.

Allosaurus is known to have been an active predator. Evidence for this comes from an *Allosaurus* tail bone that was punctured by a *Stegosaurus* tail spike, suggesting the two were engaged in a fight, as well as *Allosaurus* bite marks that were found on a *Stegosaurus* neck plate. Unlike *Tyrannosaurus*, *Allosaurus*' bites weren't strong enough to crush bone. Instead, it may have made slashing bites at its prey in an attempt to weaken it.

1

Spinosauridae

This group of theropods is one of the most intriguing. *Spinosaurus*, after which the group is named, was the largest land predator ever. However, unlike other big predators, the skulls of the Spinosauridae were long and narrow and full of conical teeth that were ideal for spearing fish. Scientists quickly realised these theropods were specialised feeders and that their main prey was aquatic. Their diet wasn't restricted to fish, however. One species, *Baryonyx walkeri*, was found with a juvenile *Iguanodon* in its stomach, along with fish scales.

Spinosaurids first appeared in the Late Jurassic and became common by the Early Cretaceous. The group is divided into two subfamilies, the Baryonychinae and the Spinosaurinae. The Spinosaurinae includes *Spinosaurus* from North Africa and the crested *Irritator* from Brazil. The *Irritator* was a particular challenge to scientists (hence the name), as fossil-poachers had tampered with its skull, lengthening its snout with plaster.

The first Baryonychinae to be discovered was *Baryonyx*, from southern England. Its

name, meaning 'heavy claw', comes from the terrifying curved claw on its first finger. This was probably used for defence, and to catch fish. This subfamily also includes *Suchomimus* ('crocodile mimic'), which was almost twice the size of the largest crocodile alive today.

--- *Key to plate* ---

1: Spinosaurus aegyptiacus
Mid-Cretaceous, Egypt
Length: 18m; Weight: 9000kg
Instantly recognisable by the huge sail on its back, which stood just over 2m high, and its incredibly long snout studded with interlocking teeth, *Spinosaurus* was built to inspire fear. Moreover, its strange and specialised form meant that it was able to carve out a niche for itself alongside the other theropods of the day,

hunting for fish and other aquatic and shore-dwelling prey, while more conventional theropods, such as *Carcharodontosaurus*, went after heftier prey on land. *Spinosaurus* would have feasted on giant coelacanths, sawfish, lungfish and sharks.

Further features point to its adaptation to a watery life. Its nostrils were positioned high on its snout, meaning it would have still been able to breathe, even when its body was

mostly submerged in water. Like early whales, it had short hindlimbs, dense bones like those of penguins, and wide, flat claws and feet which would have been perfect for paddling.

Palaeontologists have long debated the purpose of *Spinosaurus*' sail. Many now believe it was used for display, to warn off predators and to attract mates. It may also have been brightly coloured, like the fins of some reptiles alive today.

1a

1b

1c

1d

Coelurosauria

The way theropods are grouped together is constantly changing as scientists come to understand them more fully and investigate new finds. Currently, Coleurosauria includes small and large theropods that are more closely related to modern birds than allosauroids or ceratosaurs.

There are three main groups: the Tyrannosauridae (giant coelurosaurs), the Ornithomimosauria (bird-mimic dinosaurs) and the Maniraptora, which include Therizinosauria (plant-eating theropods), Alvarezsauridae (bug-eating theropods), Troodontidae (small, bird-like theropods), Dromaeosauridae ('raptor' dinosaurs) and birds.

Feathers, or at the very least 'dinofuzz', have now been found in every branch of the coelurosaur family tree and many scientists believe that all species were feathered, at least at some point in their lives.

It is this group of dinosaurs that shows us the fascinating evolution of feathers, from single, hair-like barbs to modern flight feathers. The discovery of complex feathers in non-flying dinosaurs also suggests that feathers evolved for many reasons other than flight.

They could have been used for camouflage, for warmth, to line nests and for display when attracting mates.

The Coelurosauria also gives us the extraordinary story of the evolution of dinosaurs into birds. This in turn has led to the huge shift in our perception of dinosaurs from lumbering, scaly green beasts, to agile, feathered creatures in a rainbow of colours.

───────────── *Key to plate* ─────────────

1: **Evolution of feathers in dinosaurs**
a) Simple, thin, hollow filaments, which first appeared over 150 million years ago as found on *Psittacosaurus*
b) Tufts of filaments, resembling a soft downy fuzz as found on *Dilong*
c) Numerous filaments sticking out from a central shaft as on *Sinornithosaurus*
d) Asymmetrical flight feathers, with shaft located off-centre as found in birds.

2: *Dilong paradoxus*
Early Cretaceous, China
Length: 2m; Weight: 10kg
A smaller, earlier relative of *Tyrannosaurus*, *Dilong* was the first tyrannosaur found with evidence of primitive hair-like structures, which resembled fluffy, insulating down. Like the other tyrannosaurs, *Dilong* had powerful jaws and teeth suited to

tearing meat. Its discovery proved that it wasn't just the small, bird-like dinosaurs that possessed feathers.

Dilong's fluffy down has led some experts to believe that all juvenile tyrannosaurs had a similar covering, which they may have shed on reaching adulthood, much like elephants and whales shed their hair in infancy.

Tyrannosauridae

The name means 'the tyrant lizards' and it's an apt one. The huge carnivores in this group – *Gorgosaurus*, *Daspletosaurus*, *Albertosaurus* and *Tyrannosaurus*, all from North America, and *Tyrannosaurus*' close cousin, *Tarbosaurus* from Asia – ruled the land during the final 20 million years of the Cretaceous. Their short, deep skulls allowed them to generate extreme bite forces, which more than made up for their tiny but strong forelimbs. Their massive, banana-shaped teeth were unlike other carnivores' – rather than razor-edged slashing blades they were more like giant, bone-crushing spikes. They all shared the typical tyrannosaurid shape – a massive head on the end of an S-shaped neck, two-digit hands, long hindlimbs and a long, heavy counterbalancing tail.

The large tyrannosaurids would have feasted on ceratopsians and hadrosaurs, although some tyrannosaur fossils show evidence of bite marks from other tyrannosaurids, suggesting they fought each other, or may even have engaged in cannibalism. Juveniles would have been capable of high running speeds, but this is unlikely for adults, as a fall for a fully grown tyrannosaur could easily prove fatal. However, adults would have been quick enough to hunt prey, and probably relied on both hunting and scavenging to find food.

The discovery of at least nine individuals of *Albertosaurus* from the same site at different stages of growth, and a group of 68 *Tarbosaurus* skeletons in the Gobi Desert, suggests that at least some tyrannosaurids were social animals, living and hunting together.

Key to plate

1: Tyrannosaurus rex

Late Cretaceous, North America
Length: 12m; Weight: 6000kg

The most famous dinosaur, first discovered in 1902, *Tyrannosaurus* has captured the imagination like no other. With 60 teeth, a powerful bite and a superb sense of smell, *Tyrannosaurus* was undoubtedly a ferocious killer. With the discovery in 2012 of the slightly smaller tyrannosaurid, *Yutyrannus huali*, with 20cm-long feathers, it now seems highly possible that *Tyrannosaurus* was also covered in some sort of feathers and, like other tyrannosaurids, may have hunted in packs.

Because it is known from so many good specimens, scientists have been able to build up a detailed picture of *Tyrannosaurus*' appearance, development and behaviour. We know that *Tyrannosaurus* went through a huge growth spurt from 13 to 17 years old and was fully grown by its early 20s. Juveniles had blade-like teeth, which later become conical, and over time its skull thickened and its body bulked out massively. However, being a *Tyrannosaurus* wasn't easy. They could live to be 30 years old but only 2 per cent of fossil finds showed tyrannosaurs achieving their full natural lifespan.

1

Ornithomimosauria

With their long strong legs, large eyes, slender necks, feathers and beaks, the ornithomimosaurs looked very similar to the modern ostrich, hence the name 'bird-mimic lizards', or as they are commonly known, 'ostrich dinosaurs'. They first appeared in the Early Cretaceous and lasted until the very end of the time of the dinosaurs. Most remains are known from North America and Asia, although fossils have also been found in Spain and South Africa.

Primitive forms, like *Pelecanimimus* and *Harpymimus*, were smaller and possessed numerous teeth, while the later, more advanced ornithomimids, such as *Gallimimus* and *Struthiomimus*, had toothless beaks. Some ornithomimids grew to huge sizes – *Gallimimus*, for example, was an impressive 8m long.

The ornithomimids were probably among the fastest dinosaurs; the largest rivalling the modern ostrich for speed. However, this ability would have been used to flee from predators, not to catch prey. Although these dinosaurs may have eaten small prey that they could swallow whole, they relied on plants as their main food source. Like modern herbivores, they had gastroliths to grind up tough plant matter in their stomachs and their long sloth-like arms were probably useful for pulling down branches and stripping leaves from trees. From the fossils record, they number as some of the most abundant dinosaurs in North America, which also suggests they were herbivores, as herbivores usually outnumber carnivores in an ecosystem. They may have lived together in flocks, at least as juveniles, possibly keeping together as protection from more predatory theropods.

Key to plate

1: **Ornithomimus edmontonicus**
Late Cretaceous, North America
Length: 3.8m; Weight: 170kg
Long thought to have been scaly, recently *Ornithomimus* specimens have been found with evidence of feathers: two adults with traces on the lower arm showing feathers like those in modern birds' wings and a juvenile with impressions of 5cm long hair-like filaments covering the back, upper legs and neck. A further specimen found in 2015 was the first to show feathers along the tail. Too heavy for flight, *Ornithomimus* may have used the long feathers for mating displays.

2: **Deinocheirus mirificus**
Late Cretaceous, Mongolia
Length: 12m; Weight: 6000kg
For 50 years, the fossil of this dinosaur consisted of no more than a pair of arms, 2.4m long with large 20cm long claws. At first it was presumed these arms belonged to a huge carnivore, but we now know they belonged to a strange, hump-backed ornithomimosaur. *Deinocheirus* had a long duck-like snout, a toothless beak, and its giant hands were used for no more terrifying purposes than gathering food and pulling down branches.

Oviraptorosauria

This group of bird-like dinosaurs are mostly known from Asia, although a few have now been found in North America. During the Late Cretaceous, the Bering land bridge connected these two continents, allowing dinosaurs to spread between them. Some scientists even consider oviraptorosaurs to be so bird-like they class them as true birds. They are characterised by their parrot-like beaks and many possessed elaborate head crests. They ranged in size from the turkey-sized *Caudipteryx* to *Gigantoraptor*, which grew to an impressive 8m long.

Primitive oviraptorosaurs, such as *Caudipteryx*, had four pairs of teeth in their beaks and *Incisivosaurus,* one of the most unusual-looking dinosaurs, had huge buckteeth at the front of its mouth. Later oviraptorosaurs, however, were toothless. Well-preserved fossils have revealed that many had wing feathers. At least four species had tails ending in a bony structure similar to a pygostyle, which in modern birds is used to support the tail feathers.

The name *Oviraptor* means 'egg thief' as originally they were thought to eat other dinosaurs' eggs after a fossil specimen was found raiding a *Protoceratops*' nest. However, on closer examination it turned out the *Oviraptor* had been sitting on its own eggs, just like a modern bird. This theory was backed up after a similar species, *Citipati osmolskae*, was found brooding on a nest of eggs.

Fossil finds suggest these dinosaurs were omnivores. An *Oviraptor* fossil has been found with the bones of a small lizard in its stomach and a *Citipati* with baby *Troodon* skulls. The inclusion of plants in their diet has been suggested by the gastroliths found in the stomach of a *Caudipteryx*, as gastroliths are used for grinding up tough plant matter.

Key to plate

1: Gigantoraptor erlianensis
Late Cretaceous, Mongolia
Length: 8m; Weight: 1400kg
This massive oviraptorosaur stood as tall as a giraffe and was 35 times bigger than *Citipati*, the next largest in the genus. Its existence explains the earlier discovery of huge 53cm-long oviraptorosaur eggs in the region.

It is the largest known beaked dinosaur and if it was feathered it would be one of the largest known feathered animals of all time.

2: Anzu wyliei
Late Cretaceous, North America
Length: 3m; Weight: 225kg
Nicknamed 'the chicken from hell' this is the most complete oviraptorosaur to be found outside Asia and has allowed scientists to build up a fuller picture of what American oviraptorosaurs looked like. The discovery also suggests the 'egg thief' name of the group is not such a misnomer – small prongs of bone found on the roof of the mouth

match those found in today's egg-eating snakes.

3: Heyuannia huangi
Late Cretaceous, China
Length: 1.5m; Weight: 20kg
A crestless oviraptorosaur, its fossils were found alongside thousands of eggs. On examination, the eggs are thought to have been blue-green in colour, which would have helped to camouflage them from predators.

Therizinosauridae

Known mostly from North America and Asia, the Therizinosauridae are among the most strange-looking dinosaurs and for decades they did nothing but puzzle scientists. Their long necks, wide bodies and four-toed hind feet resembled those of primitive sauropodomorphs. However, closer examination of the wrist and hip bones revealed that therizinosaurs were actually strange theropods – albeit herbivorous ones. With heavy bodies, pot bellies and short legs, they would have been too slow to catch prey. Moreover, their jaws were lined with small leaf-shaped teeth, well-suited to munching on foliage, and their rounded beaks would have been perfect for cropping leaves. But this left the mystery of their mighty arms and claws. *Therizonosaurus*, the largest of the group, had 2m-long arms tipped with 90cm-long sickle-shaped claws, and holds the record for the longest claws of all time. If this beast wasn't a meat eater, then what were the claws for?

Scientists now think members of the Therizinosauridae used their curved claws to grasp and shear leaves from branches, much like ground sloths do today. But that might not have been all they were used for – *Therizinosaurus*, for example, had to share its habitat with *Tarbosaurus bataar*, a close relative of *Tyrannosaurus*, so slashing curved claws could have proved invaluable as defence.

It is possible that they were also social animals. Three hundred *Falcarius*, a very primitive species, were found together in one place, and in 2011, a nesting ground with 17 clutches of therizinosaur eggs was uncovered. The discovery of so many eggs in one place suggests that therizinosaurs came together, at the very least for the nesting season.

Key to plate

1: Therizinosaurus cheloniformis
Late Cretaceous, Mongolia
Length: 10m; Weight: 5000kg
This species is known only from an incomplete skeleton, including parts of the forelimbs, hindlimbs, flattened ribs and its mighty arms and claws.

First discovered in 1948, in the Nemegt Formation of southwest Mongolia, the remains were initially thought to come from a turtle-like lizard and its name means 'scythe lizard, turtle-formed'. By 1970 it had been classed as a dinosaur, but it was only with the discovery of another Late Cretaceous therizinosaur, the much smaller *Segnosaurus* in 1973, that scientists were able to build up a picture of what *Therizinosaurus* may have looked like. *Segnosaurus*' fossil remains included a skull and leaf-shaped teeth, giving the first lead that *Therizinosaurus* was a herbivore. Then in 1996, skin impressions from *Beipiaosaurus*, a more primitive type, indicated that therizinosaurs were covered with a coat of long, downy feather-like fibres.

1

Troodontidae

These dinosaurs were long-legged, fast-running, bird-like carnivorous theropods. They were probably fairly common in the northern hemisphere, with fossils uncovered from as far north as Alaska, as far south as Wyoming and as far east as Mongolia. Species in this genus include the duck-sized *Mei long* from Early Cretaceous China, and *Saurornithoides mongoliensis* and *Borogovia gracilicrus*, both from the Late Cretaceous Gobi Desert. The largest species found so far was in Alaska, where troodontids grew up to 4m long.

Troodontids were adept and skilful hunters. They had acute hearing, with one ear a little higher than the other, a feature only otherwise seen in owls. This would have helped them pinpoint prey at night by hearing alone. They also had large, forward-facing eyes, giving them binocular vision, perfect for focusing and tracking down prey, even in dim light. Troodontids may have attacked young hadrosaurs as they slept – juvenile *Edmontosaurus* fossils have been found with *Troodon* bite marks. It is also possible that they hunted in packs, which would have helped them to bring down larger prey. Small mammals, lizards and insects are also likely to have made up a large part of their diet. They would probably have used their finger claws for grabbing prey and for scratching the soil in the hunt for insects.

Troodontids are also notable for their intelligence. There is a broad link between an animal's brain compared to the size of its body, and its level of intelligence. Troodontids had one of the largest brains compared to body size – about 1/100th the weight of its body, which is comparable to many modern mammals.

Key to plate

1: Troodon formosus

Late Cretaceous, North America
Length: 2m; Weight: 50kg

At first, Troodon was known only from a single, recurved (bending backwards), serrated tooth found in Montana, USA, in 1856. This was reflected in this dinosaur's name, which means 'wounding tooth'. Although it is one of the first dinosaur finds in North America, at first the tooth was thought to have come from a lizard. However, it was later established as belonging to a maniraptoran dinosaur, a clade of dinosaurs which are thought to be the ancestors of modern birds.

Troodon lived during the Late Cretaceous period, 66 to 75 million years ago. Like other troodontids, it had long, slender hindlimbs and large, retractable sickle-shaped claws on the second toes, which it raised off the ground while running.

In 1984, a fossilised *Troodon* nest was discovered by palaeontologist Jack Horner, containing a clutch of 19 eggs. The nest was dish-shaped and built from sediment. Amazingly, each egg had been preserved containing a tiny skeleton. These were the very first dinosaur embryos to have been found anywhere in the world.

1

Dromaeosauridae

A family of fast-running, agile, feathered carnivores, dromaeosaurids first appeared in the Middle Jurassic but really flourished in the Cretaceous period, when they spread across the globe. They all shared a characteristic S-shaped neck and, like other maniraptorans (the group of most bird-like dinosaurs), they had long arms that in some species could be folded against the body, as in the wings of modern birds. Dromaeosaurids also had relatively large, grasping hands, with three long fingers ending in long claws.

Within Dromaeosauridae is a subgroup, Eudromaeosauria or 'true dromaeosaurs': the 'raptors' of popular imagination. They were larger than other dromaeosaurids and the largest, *Utahraptor*, from the USA, grew to be the same size as a polar bear. The eudromaeosaurs fed mainly on vertebrate prey and, like the troodontids, possessed sickle-shaped claws on their second toes. These lethal claws might have been kept raised when running, and used as killing weapons during attacks on much larger dinosaurs – the raptors may have leapt at their prey, slashing at the flesh with their blade-like claws, or used them as crampons for clinging on to them.

A discovery in 2001 shed further light on possible dromaeosaur behaviour. Up to six individual *Utahraptor* fossils, including a baby, an adult and juveniles, were found together with the remains of plant-eating iguanodonts in a block of sandstone. It has been suggested that the *Utahraptors* attempted to attack their iguanodont prey mired in quicksand and were then trapped themselves. If this is correct, it provides further evidence that dromaeosaurs were pack-hunting killers.

Key to plate

1: Bambiraptor feinbergi
Late Cretaceous, USA
Length: 1m; Weight: 2kg
The first specimen of this small dromaeosaurid was a juvenile and because of its young age it was named after the Disney character 'Bambi'. It had long hindlimbs, suggesting it may have been a fast runner, and a large brain for its body size.

2: Deinonychus antirrhopus
Mid-Cretaceous, USA
Length: 3.4m; Weight: 73kg
Its name, meaning 'terrible claw', refers to its sickle-shaped killing claw, which it used to slice open its prey.

3: Dakotaraptor steini
Late Cretaceous, USA
Length: 5.5m; Weight: 200kg
A giant raptor, second only to *Utahraptor* in size. It had flight feathers on its arms, which it may have used to stabilise itself whilst pinning down prey. Although *Dakotaraptor* was too heavy to fly, its feathers suggest it may have evolved from a dinosaur that was capable of flight. It lived at the same time and place as *Tyrannosaurus*. If *Dakotoraptors* were pack hunters, there's a chance they were competing with young tyrannosaurs for food.

Dino Birds

In Liaoning Province, northeastern China, amid rolling farmland, lies one of the world's most fascinating fossil beds. During the Early Cretaceous period, 130 to 110 million years ago, frequent volcanic eruptions covered the animals and plants that lived there with fine particles of ash and mud. This preserved them to an extraordinary degree of detail, from soft body parts and stomach contents to skin and feathers.

Liaoning became famous in the 1990s when it reignited the debate about the relationship between birds and dinosaurs with a wealth of new feathered finds. The feathers found here range from the very simple to those found on modern birds, revealing that modern feathers (and wings) evolved long before the ability to fly.

These 'dino birds', as they became known, lived in Liaoning's warm climate, in a lush forest interspersed with lakes. They lived alongside giant insects, shrew-like creatures and mammals the size of dogs, which preyed on smaller dinosaurs. There were frogs and turtles, similar to those around today. Primitive birds flitted from branch to branch while strange feathered dinosaurs climbed the trees.

Key to plate

1: **Sinornithosaurus millenii**
Early Cretaceous, China
Length: 90cm; Weight: 1.5kg
This little feathered dromaeosaurid was closely related to *Velociraptor* and may be the first known venomous dinosaur. Scientists think they have found grooves in its fang-like teeth that led to what may have been venom sacs in its jaw. Researchers have described it as a predator of small birds and dinosaurs, that would have swooped down on its prey from low-hanging tree branches in order to attack unseen.

2: **Confuciusornis sanctus**
Early Cretaceous, China
Length: 50cm; Weight: 1kg
Around the size of a pigeon, *Confuciusornis* was an evolutionary link between dinosaurs and birds. It is the earliest known bird to abandon its reptilian teeth in favour of a lightweight keratin beak. It could certainly fly but unlike modern birds, still had clawed fingers on its wings. The feet, with their curved claws, were adapted for perching.

3: **Mei long**
Early Cretaceous, China
Length: 40cm; Weight: 0.4kg
This duck-sized troodontid is known for its sleeping posture after a fossil was found with its beak tucked under its wing, just like a roosting bird. Its name means 'sleeping dragon'.

4: **Sinosauropteryx prima**
Early Cretaceous, China
Length: 1.7m; Weight: 0.55kg
In 1996, this became the first non-bird dinosaur found with feathers. It was covered with a coat of very simple, reddish, down-like filaments known as 'protofeathers', thought to be a primitive type of feather. Because *Sinosauropteryx* was only distantly related to birds, the discovery of its protofeathers suggests that many theropod dinosaurs were feathered rather than scaly, as previously thought.

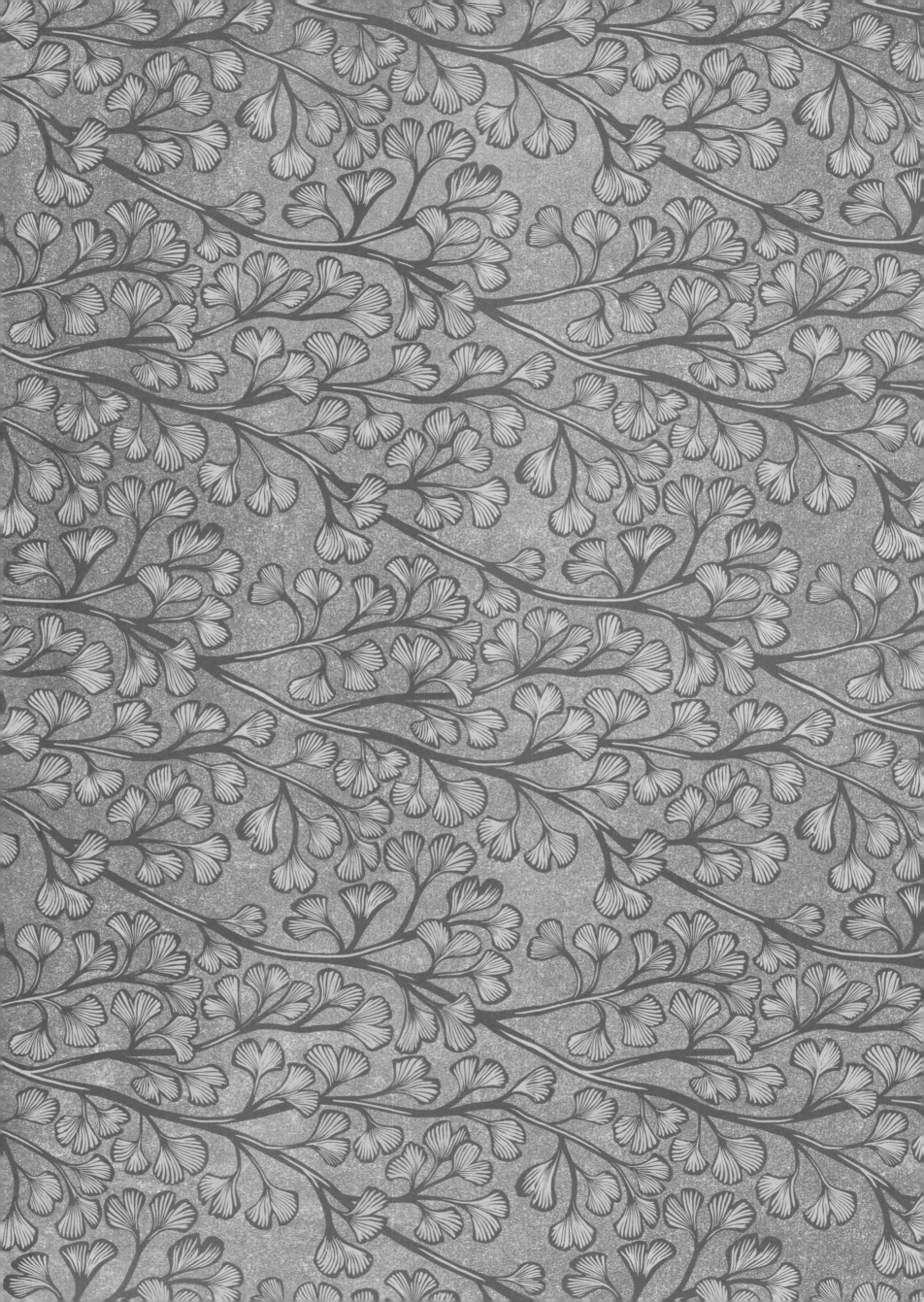

Gallery 3

Ornithopoda

Ornithopoda

The ornithopods were a diverse group of ornithischian ('bird-hipped') dinosaurs that included some of the most successful herbivores of the Mesozoic. The name 'Ornithopoda' means 'bird feet', referring to their three-toed feet, although many early forms had four toes. Ornithopods were characterised by a horny beak and a lack of armour plating, as found on other ornithischians.

Known from a range of habitats, they first appeared in the late Middle Jurassic and lasted until the end of the Cretaceous. They included the hetorodontosaurids, hypsilophodonts, iguanodonts and hadrosaurs.

Early ornithopods included groups such as the hetorodontosaurids and hypsilophodonts. They tended to be small and swift, moving on two legs. Later ornithopods were larger and better adapted to grazing on all fours, although they never rivalled the sauropods for size. Unlike their earlier relatives, ornithopods also tended to lose their front teeth and developed cheek pouches for processing food.

One of the secrets of their success was their powerful teeth and jaws, which became

remarkably well-adapted to chewing. The cheek teeth were arranged in such a way as to be able to grind up food to a finer degree than most dinosaurs. This would have allowed them to gain quicker access to nutrients from their food, and they may well have been the most efficient herbivores in reptile history. Their fossils have been found on every continent and, by the end of the Cretaceous, they counted as some of the most abundant species of dinosaur.

--- *Key to plate* ---

1: *Tenontosaurus tilletti* **(being attacked by a pack of** *Deinonychus***)**
Early Cretaceous, North America
Length: 8m; Weight: 1500kg
A close relative of *Iguanodon*, this ornithopod would have been capable of walking both on all fours or on its hind legs. It had a stiff and bony tail, which made up over half of its body length.

It is one of the few dinosaurs to be discovered with medullary bone tissue, a material inside the bone, found only in females, that stores calcium for the production of eggs. It is also found in modern birds and in *Tyrannosaurus* and *Allosaurus*. As *Tenontosaurus* is only distantly related to these theropods, it suggests all dinosaurs produced medullary bone tissue. Moreover, it was found in a specimen that wasn't yet fully grown, which suggests

dinosaurs reached sexual maturity before they finished growing.

Many *Tenontosaurus* fossils have been found with those of the bird-like theropod, *Deinonychus*, and a juvenile *Tenontosaurus* has even been found with *Deinonychus* bite marks in its bones. Although it is unknown whether *Tenontosaurus* was alive when the bite marks were made, or if the *Deinonychus* was scavenging.

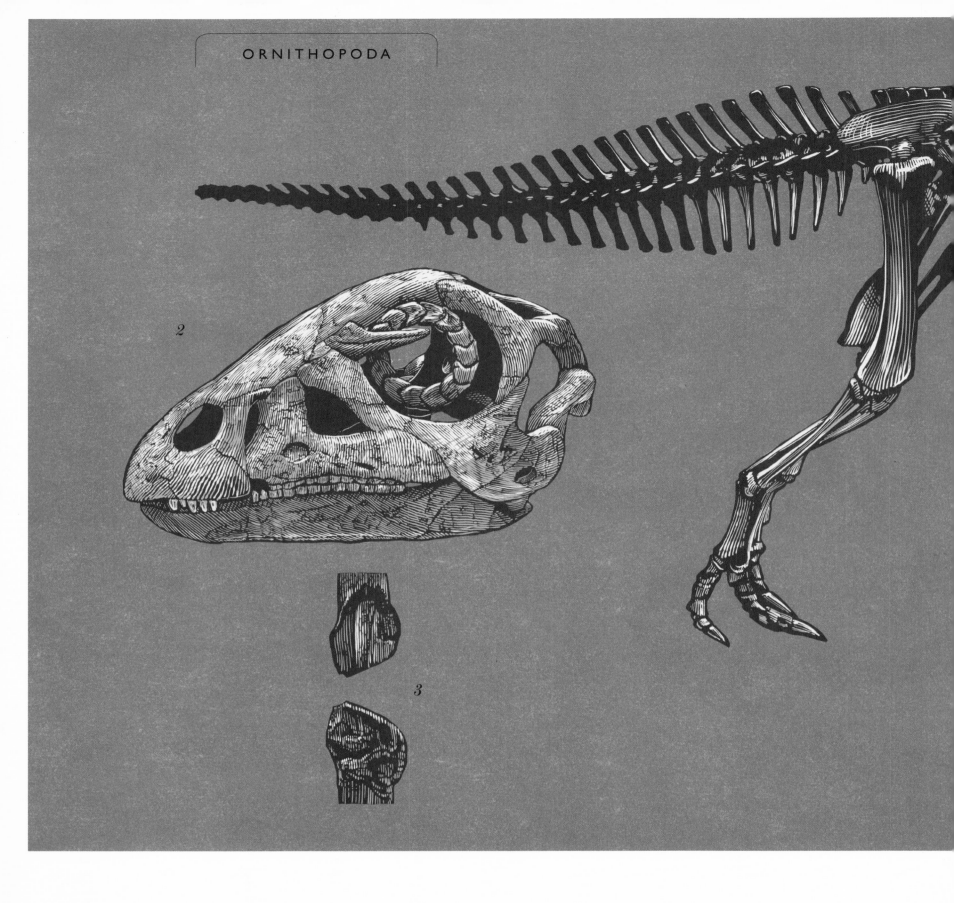

Primitive ornithopoda

These dinosaurs were fast, light herbivores that ran around on two legs. Primitive ornithopods were also characterised by their small size, sharp beaks, cheek pouches and long tails, stiffened by bony rods.

They lasted from the Late Jurassic to the end of the Cretaceous period and are known from a number of species, including *Hypsilophodon*, *Atlascopcosaurus*, *Oryctodromeus* and *Orodromeus*.

Some primitive ornithopods are believed to have been burrowers. *Oryctodromeus* was the first dinosaur to have been discovered in a burrow about 50cm underground. Three individuals were found together – an adult and two juveniles, suggesting the burrow was used to raise and protect young. Both *Oryctodromeus* and *Orodromeus* also had broad snouts and forelimbs adapted to digging.

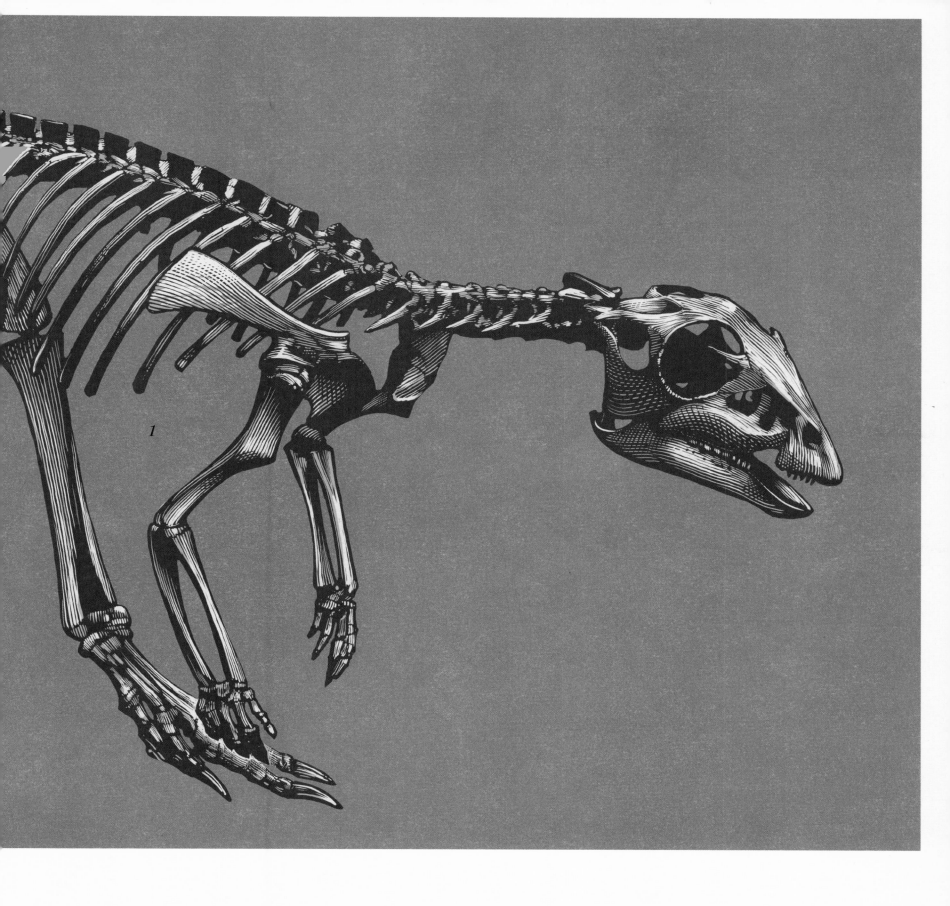

1: Hypsilophodon foxii

Early Cretaceous, Isle of Wight, England

Length: 1.8m; Weight: 20kg

First discovered in 1849, for many years *Hypsilophodon* was wrongly believed to have had opposable toes and was frequently depicted as a tree climbing dinosaur. It is now known to have lived on the ground, where it browsed on low-growing vegetation. Its beak may also have been well-adapted for tugging seeds from cycad cones, which it could then have crunched up with its broad back teeth.

The shape of its skull and jaw suggests that it had cheek-like structures, which would have helped it to chew its food before swallowing. It also possessed four toes on each back foot (unlike most other ornithopods, which had three) and jutting, bony eyelids.

All *Hypsilophodon* fossils found so far have come from the Isle of Wight, in southern England. Many bones have been found together in one area and it's thought the individuals died in quicksand. The fact they were found together also suggests these dinosaurs lived in herds.

2: Hypsilophodon skull

The skull had large eye sockets and a sclerotic ring – a collection of small bones around the eye, which is thought to have strengthened the eyeball and helped with focusing, particularly in dim light.

3: Rear teeth

The rear teeth were ridged, broad and chisel-shaped with a cutting edge, and were well-adapted to stripping leaves from vegetation.

The Jurassic Period

A mass extinction marks the boundary between the end of the Triassic and beginning of the Jurassic. At least half of all known species became extinct, leaving a niche for the dinosaurs, who became the dominant land animals.

This period also saw the continued breakup of the vast supercontinent, Pangaea. Oceans flooded the gaps between the continents and sea levels rose, producing a mild, wet climate. Plant life flourished on Earth and large areas were covered in forests of ferns, ginkgoes, conifers and cycads. Early mammals crept through the foliage while huge sauropods roamed the land, feasting on the vegetation, along with stegosaurs, armoured ankylosaurs and the smaller ornithopods.

Many of the Jurassic theropods, such as *Allosaurus*, were huge; capable of killing even the largest sauropods. There were also the agile coelurosaurs and earliest-known bird, *Archaeopteryx*, most likely descended from an early coelurosaurian dinosaur. These early birds would have shared the skies with the leathery-winged pterosaurs.

Fish-like ichthyosaurs swam in the seas, along with the plesiosaurs, giant prehistoric crocodiles, sharks and rays. Squid-like cephalopods, ammonites, sponges and molluscs also flourished in the Jurassic seas.

1: Tianyulong confuciusi
Length: 70cm; Weight: 800g
This cat-sized dinosaur is famous for being covered in feather-like structures along its back and tail, forming a kind of fuzz. Before its discovery, nearly all dinosaurs known to have feathers were the bird-like coelurosaurs from the saurischian branch. *Tianyulong*, however, was an ornithischian dinosaur. The presence of bristles suggests either that feather-like structures evolved independently in ornithischians or that the bristles came from a common ancestor of ornthischians and saurischians. Either way, it hints that dinosaurs had a much more varied array of body coverings than was previously thought.

2: Mongolarachne jurassica
Body length: 1.65 cm;
Leg length: 5.82 cm
The largest-known fossil spiders, covered in feathery hair. They made orb-shaped webs with an incredibly sticky silk.

3: Dicksonia
A genus of tree fern, with a stout, erect, fibrous trunk, and large fronds at the top.

4: Williamsonia
This seed plant had a sturdy stem and many fern-like leaves, and produced 10cm-long flowers. *Williamsonia* first appeared in the Triassic but became abundant in the Jurassic period.

5: Ginkgo
A non-flowering plant with sprays of stalks, each with a single seed and lobed leaves. A single species, *Ginkgo biloba*, survives today in the wild in China.

6: Juramaia sinensis
Length: 7–10cm; Weight: 15g
A small shrew-like mammal that lived in China 160 million years ago. It is also the earliest known ancestor of placental mammals (the large group of modern mammals, including humans, with a placenta for nourishing their unborn young). Its forelimbs were adapted to climbing and it would have scampered through the fern fronds, hunting for insects.

1

Iguanodon

One of the first dinosaurs to be formally named, *Iguanodon*, along with *Megalosaurus* and *Hylaeosaurus*, was also one of the three dinosaurs originally used by Sir Richard Owen to define the group Dinosauria, or 'terrible lizards', in 1842.

The first discovery dates back to 1822, when a scattering of teeth were unearthed in Sussex, England. Three years later, Gideon Mantell, a fossil enthusiast and country doctor, described them as belonging to a creature he named *Iguanodon*, because of the similarity of the teeth to an iguana's, only much larger. Based on the size of the teeth, it was thought that *Iguanodon* was 20m long (twice the size we now know it to be), and like an iguana, it was imagined as a lumbering four-legged beast. Early re-creations also placed *Iguanodon*'s thumb spike on the tip of its nose, as seen in the sculptures built in the early 1850s in Crystal Palace, London.

Then, in 1878, the perception of *Iguanodon* changed with the discovery of more than 30 complete skeletons in a coal mine in Bernissart, Belgium. The animals are thought to have been the victim of a flash flood. Many of the specimens were nearly

complete and some were also jointed, showing how the bones fitted together. Over time *Iguanodon* would become one of the world's most studied dinosaurs.

Recent discoveries, and re-examination of old *Iguanodon* bones, have shown that iguanodontids are much more diverse than was previously thought.

—————————————— *Key to plate* ——————————————

1: Iguanodon bernissartensis
Early Cretaceous, England, Belgium and Germany
Length: 10m; Weight: 3200kg
This large, bulky herbivore probably spent most of its time walking on all fours, although it could also move around on its hind legs, which were longer than its forelimbs. It was built for walking, not running, and would

have browsed low to the ground. It had a bony beak and closely packed teeth, well-adapted for grinding up tough plant matter. The shape of its skull suggests that, like other ornithopods, it had a cheek-like structure to hold food in its mouth.

2: Iguanodon **hand**
As shown here, *Iguanodon* had three

middle fingers on each hand and a grasping little finger used for foraging for food. The purpose of the thumb spike, one of *Iguanodon*'s best-known features, is still being debated, with some scientists proposing it was a defensive weapon against predators and rival *Iguanodons* and others claiming it was used to break open large seeds and fruit.

Hadrosauridae

For millions of years this group of 'duck-billed dinosaurs' (so-called because of their flat, toothless duck-like bills) were the world's dominant herbivores. They evolved from the iguanodontids and spread throughout Late Cretaceous Asia, America and Europe. They were the cows of their day, moving in large herds across the Mesozoic landscape, stripping back low-growing vegetation with their rapidly replaced teeth.

Their beaks were ideal for clipping off leaves and all species of hadrosaur had hundreds of small cheek-teeth, known as a dental battery, for grinding up plant matter. They also possessed a unique eating motion, not seen in any animal alive today, made possible by a special hinge between the upper jaws and the rest of the skull. When chewing, their teeth slid sideways across each other, which allowed for very precise grinding and slicing of plant material, and put the least possible pressure on the rest of the skull.

There are two subfamilies of hadrosaurs: the Lambeosaurinae, which had hollow crests, and the Saurolophinae, which had solid crests or none at all.

Key to plate

1: Parasaurolophus walkeri
Late Cretaceous, North America
Length: 9m; Weight: 2500kg
Scientists have made a computer model of this hadrosaur's skull and recreated the sounds it would have made by blowing air through the twisting tubes inside the crest. The results were low, booming notes that would have echoed across the plains.

2: Tsintaosaurus spinorhinus
Late Cretaceous, China
Length: 10m; Weight: 3000kg
With its forward-facing crest, this elephant-sized herbivore is also known as the 'unicorn dinosaur'. The crest was hollow and may either have helped it to regulate temperature, enhance its sense of smell or could have been used purely for display.

3: Lambeosaurus lambei
Late Cretaceous, Canada
Length: 9m; Weight: 2500kg
Closely related and similar to *Corythosaurus*, but with a crest that shifted forwards and with hollow nasal passages at the front. The shape of the crests are known to have changed as the animal aged.

4: Edmontosaurus regalis
Late Cretaceous, North America
Length: 14m; Weight: 4000kg
For many years thought to have been crestless, until 2013 when it was discovered that this hadrosaur had a soft, supple wattle or comb on its head, like that of a cockerel, 20cm tall. This was most likely used for display.

5: Saurolophus angustirostris
Late Cretaceous, Mongolia
Length: 12m; Weight: 3500kg
The most common Asian hadrosaurid, its long spike-like crest was made up entirely of its nasal bone, with internal chambers. The crest was smaller in juveniles.

6: Corythosaurus casuarius
Late Cretaceous, Canada
Length: 9m; Weight: 2500kg
Like *Parasaurolophus*, *Corythosaurus* is thought to have used its crest for display and vocalisation, with the hollow tubes inside working as a resonance chamber.

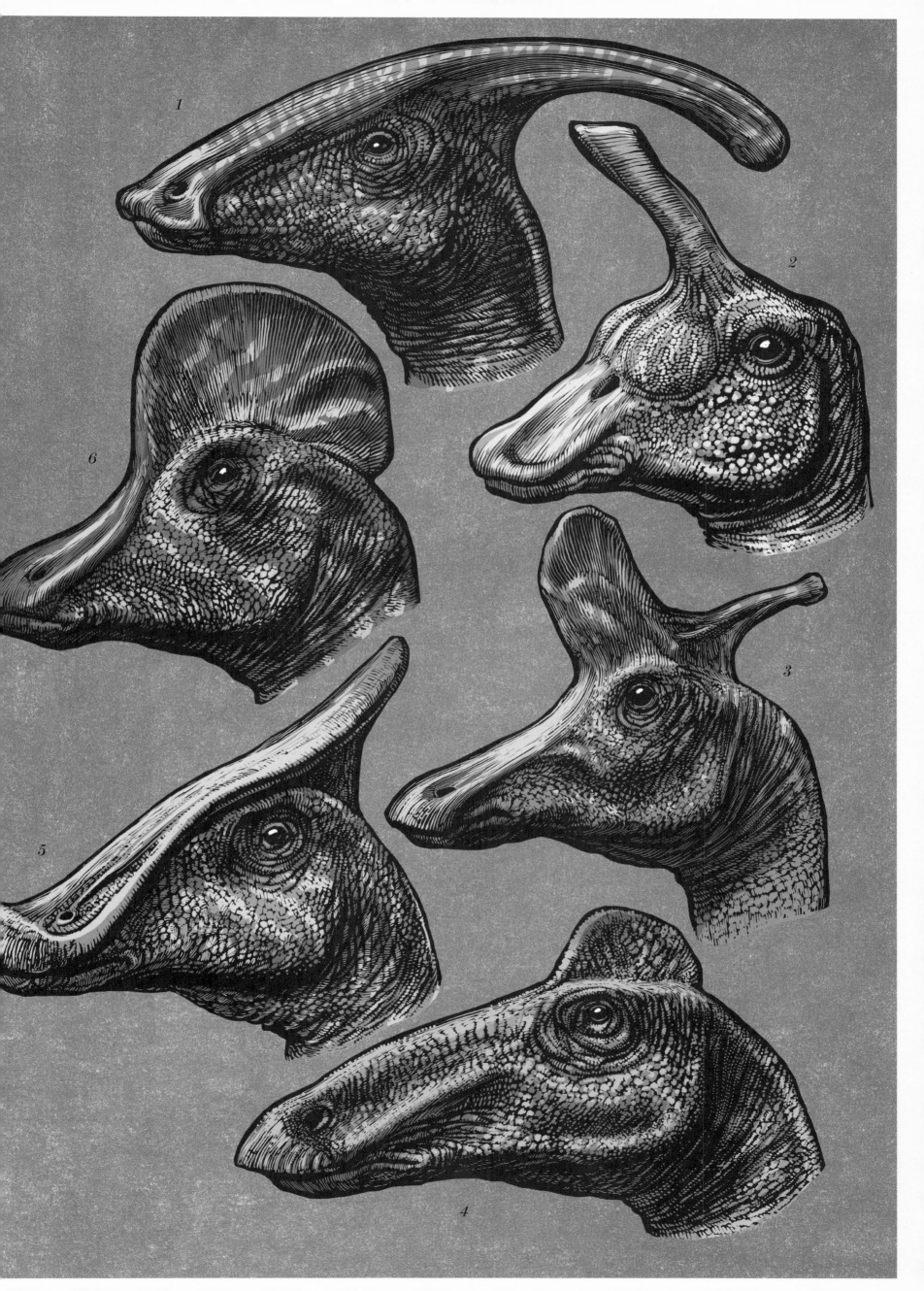

Egg Mountain

Around 77 million years ago, on what is now a plateau in the Rocky Mountains of Montana, USA, a herd of *Maiasaura* came to make their nests and lay their eggs. Many would never hatch. Instead, they were covered by volcanic ash, preserving them for future study.

This fossil site, known as 'Egg Mountain', was excavated in 1979 by palaeontologist Jack Horner. It revealed for the first time that dinosaurs cared for their young. The site contained hundreds of *Maiasaura* specimens, from adults through to juveniles and babies, along with bowl-shaped nests filled with eggs. The presence of young up to two months old near the nests suggested that *Maiasaura* cared for its offspring for much longer than other dinosaurs. Regurgitated plant matter was also found at the site, indicating that parents were bringing food to their young. This was also evidenced by partly worn teeth on the hatchlings, who were too weak to find food for themselves.

Maiasaura was not the only dinosaur found at Egg Mountain. Among the fossils were specimens of *Troodon*, who also used the area as a nesting site, and the remains of one of the largest flying reptiles, an as yet unnamed pterosaur.

Key to plate

1: Maiasaura peeblesorum
Late Cretaceous, USA
Length: 9m; Weight: 3000kg
Its name means 'good mother lizard' because it was found caring for its eggs and young. A duck-billed dinosaur, or hadrosaur, *Maiasaura* was a herbivore, feeding on plants, leaves, berries and rotting wood. It is thought to have roamed the Cretaceous plains in vast herds, 10,000 strong, before returning to its nesting sites.

It is possible that *Maiasaura* raised their offspring to the point that they could keep up with the herd and then migrated across the dry plains in search of food.

2: Nest
Diameter: 1.8m
The crater-shaped nests were made of earth and contained rotting vegetation. Too heavy to sit on the eggs themselves, the *Maiasaura* would have used the heat from the rotting vegetation to incubate their clutch. Each nest was around 7m apart, less than the length of an adult *Maiasaura*, making up a vast dinosaur-style maternity ward.

Horner also discovered that the fossilised nests were layered one on top of another in the rock, suggesting herds of *Maiasaura* returned to the same nesting site each season.

3: Eggs
Length: 15cm
The eggs were about the size of a grapefruit, with around 30 in each nest, laid in a circular or spiral pattern.

4: Hatchling
Length: 40cm; Weight: 1kg
Tiny at birth, hatchlings grew fast – 147cm in their first year, and reached adult size at the age of 8 years. *Maiasaura* had an 89.9 per cent mortality rate in their first year of life, dropping to 12.7 per cent for the second year. They would have walked on two legs at first, though as adults they moved mostly on four.

Gallery 4

Thyreophora

Thyreophora

Stegosauria

Ankylosauria

The Cretaceous Period

Thyreophora

This group of ornithischian dinosaurs lived from the Early Jurassic to the Late Cretaceous period. Their fossils are known from all over the globe and from a diverse range of habitats. The name 'thyreophorans' means 'shield bearers', but they are commonly known as 'armoured dinosaurs' to reflect their tough, bony body armour, which they developed to protect themselves from the predators of the day.

Primitive thyreophorans were smaller and had much less body armour than more developed forms. *Scelidosaurus*, for example, a primitive thyreophoran of Early Jurassic England, was 4m long and lightly covered with bony deposits, embedded in the skin, like those of crocodiles. These deposits, known as osteoderms, were in turn covered in keratin (the same material fingernails and horns are made from) and arranged in rows along the animal's back. The osteoderms were sharp enough and strong enough to break a predator's teeth.

Thyreophorans are divided into two major groups – the Stegosauria, with their rows of plates and spines, and the more heavily armoured, tank-like Ankylosauria.

All thyreophorans were herbivores. They had hoof-like claws and are thought to have possessed a horny beak covering the tips of the upper and lower jaw, which would have been used to clip leaves from low-growing branches. Unlike other ornithischians such as the hadrosaurs, they didn't go on to develop a sophisticated dental battery of chewing teeth. Instead, their teeth were designed for slicing and cutting. They processed their diet of leaves, twigs and other tough plant matter in their large guts.

Key to plate

1: Scutellosaurus lawleri
Early Jurassic, USA
Length: 1.3m; Weight: 10kg
The most primitive thyreophoran, and one of the earliest-known ornithischians, *Scutellosaurus* was discovered in red claystone in Arizona, in sediment dated to around 196 million years ago. *Scutellosaurus* is known from a small portion of its skull, teeth and several partial skeletons.

Scutellosaurus walked on two legs (almost all other thyreophorans walked on four) although it may have foraged on all fours. It was swift and agile, with a long tail for balance. However, it didn't have to rely on speed alone to escape from predators, such as *Coelophysis*. Its name, meaning 'small shield lizard', reflects the fact it was covered in hundreds of bony scutes. *Scutellosaurus* had more than 300 of

these studs, ranging from small lumps on its back to ridged plates, like tiny versions of those found on *Stegosaurus*.

Scutellosaurus had leaf-shaped teeth with serrated edges, which it used to snip leaves from the lower branches of trees and bushes. The lack of wear on the teeth suggests that *Scutellosaurus* swallowed its food without chewing it.

1

Stegosauria

The stegosaurs were striking-looking dinosaurs with bodies covered in spikes and rows of plates along their backs. Like other thyreophorans, they also had small bony deposits in their skin, mostly around the neck and hips.

Some stegosaurs had spikes on their sides, which would have been a deterrent to many predators, and fossil evidence suggests that their tail spikes were actively used as defensive weapons. Tail spikes have been found with damaged tips, along with theropod bones with puncture wounds that fit the shape of the spikes.

The function of stegosaur plates has been hotly debated. It has been suggested that they were covered in blood vessels and used to regulate temperature by radiating heat away from the body, but it is now thought they were mainly used for identification and in courtship displays. The blood vessels pumping around the plates would have enabled the animals to blush red to attract a mate or warn off rivals.

Early stegosaurs were only about 2.5–3m long, but later forms could reach up to 9m or more. Their narrow snouts suggest that they were selective eaters, carefully choosing

which plants to eat. They may also have been able to rear up on their hind legs when browsing for food.

Stegosaurs were at their most common, and most diverse, in the Middle and Late Jurassic and had died out by the Early Cretaceous period. They lived all over the globe, although most fossils have been found in North America and China.

───────────────────────── *Key to plate* ─────────────────────────

1: Kentrosaurus aethiopicus
Late Jurassic, Tanzania
Length: 5m; Weight: 1000kg
The spikiest of the stegosaurs, *Kentrosaurus* had long spikes on its shoulders to protect itself from side attacks. It could also swing its tail in a 180 degree arc, at high speeds, making it capable of causing serious damage to would-be attackers.

2: Huayangosaurus taibaii
Middle Jurassic, China
Length: 4m; Weight: 850kg
One of the smallest and most primitive stegosaurs, *Huayangosaurus* had a broader skull than later stegosaurs and teeth at the front of its mouth, as well as relatively long forelimbs. It also shared some similar features with ankylosaurs, suggesting it may have lived at a time when stegosaurs and ankylosaurs were diverging along separate evolutionary lines.

3: Stegosaurus armatus
Late Jurassic, USA and Portugal
Length: 9m; Weight: 2300kg
The largest of the stegosaurs, it had diamond-shaped plates along its arched back and a long, narrow skull. Its forelimbs were much shorter than its hind legs, which suggests it wouldn't have been able to move very fast, as when running, its back legs would have tripped up its front legs. It is thought to have eaten low-growing vegetation, including ferns, mosses and cycads.

1

Ankylosauria

The Ankylosauria were the most heavily armoured dinosaurs ever to walk the Earth, with large parts of their bodies covered in bony plates. There were two main groups – the nodosaurs and the ankylosaurs. Both had plates and small spikes embedded in their skin, but nodosaurs also had long spines on the sides of their bodies, supported by a specialised knob of bone on each shoulder blade. The spear-like spines were most likely used as self-defence against predators, although they may also have been used against others of the same species, to fight over mates and territory.

Ankylosaurs lacked the spikes of the nodosaurs and were generally wider, with shorter, broader snouts, suggesting they were more generalised feeders. They also had broad, triangular, armoured heads and some species had a further impressive weapon against predators – a club at the end of the tail. These clubs were made of several plates of bone fused together with soft tissue and could be swung at theropods with bone-shattering force.

Given their weight and sturdy limbs, both nodosaurs and ankylosaurs were probably slow-moving animals that scoured the ground for low-growing vegetation to feed on. They didn't chew their food but slowly digested it in their massive guts.

The Ankylosauria lived from the Jurassic until the very end of the Cretaceous period and their fossils have been found on every continent except Africa.

Key to plate

1: Euoplocephalus tutus
Late Cretaceous, Canada
Length: 6m; Weight: 2500kg
Euoplocephalus was covered in bony armour, except for parts of the limbs and tail. Its tail club was supported by bony tendons, but it couldn't lift its club far from the ground. Powerful muscles near the base of the tail enabled it to swing the club from side to side, most likely at the fragile shin bones of advancing predators.

2: Ankylosaurus magniventris
Late Cretaceous, USA and Canada
Length: 7m; Weight: 3000kg
The largest of the ankylosaurids, with horns on its head, a beak and small, leaf-shaped teeth. Even its eyelids were covered in bony plates. Some of its plates were fused together, giving its armour-plating extra strength: it would have needed it to see off the predators of the day, which included *Tyrannosaurus rex*.

3: Sauropelta edwardsorum
Early Cretaceous, USA
Length: 5m; Weight: 1500kg
A nodosaur with enormous neck spines to protect it against predators as well as a covering of body armour, including small bony nodules and parallel rows of domed scutes along its neck and back. It lived in wide floodplains and possibly in herds, as the fossils of at least five animals have been found together.

The Cretaceous Period

The Cretaceous was the last and longest period in the Mesozoic era. It followed on from a major extinction at the end of the Jurassic and was a time of huge change. Flowering plants appeared for the first time and rapidly diversified, spread with the help of pollinating insects including bees, wasps, ants, beetles and butterflies.

There were more dinosaur species in the Cretaceous than at any other time. In the north, vast herds of horned ceratopsians browsed on vegetation, alongside the armoured ankylosaurs. Iguanodontids spread everywhere except Antarctica, while mighty titanosaurs roamed the southern continents. Theropods were still the apex land predators.

In the skies, the pterosaurs faced competition from many different species of birds and the ancestors to modern birds appeared for the first time. Icthyosaurs, plesiosaurs and giant mosasaurs swam in the seas along with sharks and modern rays. Early frogs, salamanders, turtles, crocodiles, small mammals and snakes thrived along the coasts.

But by the end of the Cretaceous some scientists believe that dinosaurs were in decline as the climate became cooler and wetter. Then 66 million years ago came the mass extinction that saw the end of the non-bird dinosaurs, as well as many other Mesozoic life forms. Life on land would never be this big again.

Key to plate

1: Minmi paravertebra
Length: 3m; Weight: 300kg
A small ankylosaur from Early Cretaceous Australia with a beak and serrated cheek teeth. Unlike most other ankylosaurs it may have been a speedy runner and lacked any armour on its head. It lived in a habitat made up of a mix of woodlands and floodplains. Studies of its gut suggest it ate the seeds and fruit of flowering plants, along with ferns and other soft plant material.

2: Muttaburrasaurus langdoni
Length: 8m; Weight: 2800kg
An ornithopod with rows of grinding teeth and a curious bump on its snout with a hollow chamber inside. This may have enhanced its sense of smell or helped it to make loud calls. Its diet probably included ferns, cycads, clubmosses and conifers.

3: Mythunga camara
Wingspan: 4.7m; Weight: uncertain
A large, primitive pterosaur with widely spaced, interlocking teeth. It lived near the coast of a large, inland sea, where it would have soared on the air currents and dived down to feed on fish.

4: Conifer forest
Conifer trees persisted from the Jurassic period and had massively diversified by the Middle Cretaceous. Conifer forests covered much of Australia in the Cretaceous, especially along the coasts. Understoreys were made up of ginkgoes, cycads, clubmosses and horsetails.

5: Flowering plants
By the Early Cretaceous, the first angiosperms (flowering plants) had began to appear. The earliest known forms include *Clavatipollenites* and the Magnolias.

6: Nanantius eos
Wingspan: 35cm; Weight: 80g
An early Cretaceous bird, about the size of a blackbird. It had clawed wings and its head and neck closely resembled that of a feathered theropod. It is thought to have fed on small fish and other tiny sea creatures. It belonged to the enantiornithines, a branch of primitive birds that was separate to the branch that led to modern 'true' birds. Enantiornithines became extinct at the end of the Cretaceous period.

Gallery 5

Marginocephalia

Marginocephalia

Pachycephalosauria

Ceratopsia

Fighting Dinosaurs

Marginocephalia

The name of this clade of dinosaurs means 'ridged heads', as they are characterised by a unique skull structure – a bony shelf or frill at the back of the skull. There were two main groups: the thick-skulled Pachycephalosauria (which included dinosaurs such as *Pachycephalosaurus* and *Stegoceras*) and the horned Ceratopsia (such as *Triceratops* and *Styracosaurus*) which also possessed a rostral bone, or beak.

The greatly thickened skull of the pachycephalosaurs formed a dome on top of their heads. They had a small ridge at the back of the skull, often covered in small, bony lumps and spikes. In ceratopsians, the ridge was generally much larger, forming a bony frill, thought to be used for display, to communicate with animals of the same species and to attract mates. Ceratopsian species were also often adorned with long horns jutting out from the top of the nose, cheeks or in some species, from the top of the frill itself.

The pachycephalosaurs and early ceratopsians walked on two legs, but all later ceratopsians walked on all fours. Marginocephalians were also characterised by simple, peg-like teeth surrounded by a horny sheath of keratin. The teeth were arranged in stacked rows, so could be easily replaced, and had serrations, which helped them slice through vegetation. Tough plant matter would have been digested in their large guts.

Marginocephalians first evolved in the late Middle Jurassic but are mostly known from the Cretaceous period. They lived in diverse environments but were quite restricted geographically, with most fossils finds coming from Asia and western North America.

Key to plate

1: *Diabloceratops eatoni*
Late Cretaceous, USA
Length: 5.5m; Weight: 2000kg
There has been a recent explosion in the discoveries of primitive ceratopsians. One of these is *Diabloceratops*. Discovered in 1998 and excavated in 2000, this relatively new species may be a primitive ancestor of the well-known ceratopsians *Triceratops* and *Styracosaurus*. Its primitive features included an

opening in its skull that disappeared in later ceratopsians.

It was medium-sized but strange-looking, with its rounded nose and a skull that bristled with horns – long horns above the eyes, a small one on the nose and two 50cm-high horns on its neck frill that rose upwards before curving sideways at the tips. Its name, meaning 'devil', is in reference to the horns on its frill.

When *Diabloceratops* was alive,

North America was split into two landmasses – Laramidia and Appalachia, and a shallow sea known as the Western Interior Seaway covered most of the midwest. *Diabloceratops* lived in Laramidia, in an area covered in lakes, floodplains and rivers, where it would have used its beaked mouth to feed on low-growing plants.

Pachycephalosauria

Most pachycephalosaurs lived during the Late Cretaceous period in what is now North America and Asia. They had thick domes on top of their skulls and a shelf at the back, dotted with small bony lumps and spikes. Some fossils have been found with flat skulls, but while these were once thought to be distinct species, they are now believed to represent juvenile specimens of existing species, as there is evidence their skulls changed dramatically as they grew, thickening around the dome. Some species had skulls up to 23cm thick (by comparison, a human skull is only 6.5mm thick).

Pachycephalosaurs are thought to have used their helmet-like heads either for display or to fight each other, most likely over mates or territory. In one study, 20 per cent of the skulls showed signs of injuries, which would support the idea they used their heads in combat. It's unclear, however, whether they fought in head-butting contests, as mountain goats and bighorn sheep do today, or by 'flanking' each other (pushing their heads into each others' soft flanks) like male giraffes.

The pachycephalosaurs were herbivores and were equipped with beaks for clipping vegetation and small, ridged teeth to help with chewing. They walked on two legs and had small, weak arms and hands.

Key to plate

1: Stegoceras validum
Late Cretaceous, North America
Length: 2m; Weight: 40kg
Goat-sized, with an S-shaped neck and a stiffened tail, Stegoceras was one of the first known pachycephalosaurs. Its head was domed and studded with knobs and spikes. It is thought to have had good binocular vision and a strong sense of smell. Its small, serrated teeth were well-suited to a varied diet of leaves, seeds, fruit and insects.

2: Dracorex hogwartsia
Late Cretaceous, USA
Length: 2.4m; Weight: 45kg
Formally described in 2006, this dinosaur's name means 'the dragon king of Hogwarts', inspired by the combination of its dragon-like appearance and the Harry Potter books. However, it's since been argued that this isn't a new species at all, but a juvenile specimen of the much larger Pachycephalosaurus.

3: Stygimoloch spinifer
Late Cretaceous, USA
Length: 3m; Weight: 77kg
Characterised by clusters of spikes on the back of the skull, it has a long horn surrounded by 2–3 smaller hornlets. Its skull is very similar to Dracorex, but with shorter hornlets and a thicker dome on its skull. Some scientists think that, just as Dracorex is a juvenile Pachycephalosaurus, Stygimoloch is the same species in near-adult form.

4: Pachycephalosaurus wyomingensis
Late Cretaceous, USA
Length: 4.5m; Weight: 450kg
The largest known pachycephalosaur with a ring of bony spikes around its extremely thick skull. No other fossil remains have been found, apart from the skull. It would have walked on its hind legs and had short arms, a bulky body and a heavy tail. It is thought to have lived on a diet of leaves, seeds and fruits.

Ceratopsia

The oldest-known ceratopsians first appeared in Asia during the Late Jurassic period, around 158 million years ago. Early forms, such as *Psittacosaurus*, walked on two legs and lacked the bony neck frill of its later relatives. By the Late Cretaceous period, ceratopsians had diversified into a family of four-legged dinosaurs with an incredible array of frills and horns on their skulls, and were among the last non-bird dinosaurs to walk the Earth.

Ceratopsians had mouths tipped with a horny beak and rows of cheek teeth well-suited to a diet of tough vegetation. The huge, heavy neck frill of later ceratopsians may have been used as armour against attack by the theropods of the day, which included *Tyrannosaurus*. Other, smaller ceratopsian frills would have provided little defence against predators, and were most likely used for signalling or as display by males, much in the same way that stags use their antlers today.

Many ceratopsians travelled in herds, as evidenced by the bones of hundreds of individuals found together in what are known as 'bone beds' in the western United States. These herds would have provided protection against predators. Ceratopsians could have stampeded to ward off attackers or circled together, protecting the old and the young at the centre of the herd, much like the behaviour of modern elephants.

--- *Key to plate* ---

1: Psittacosaurus mongoliensis
Early Cretaceous, China, Mongolia and Russia
Length: 1.5m; Weight: 15kg
Psittacosaurus had a pair of stubby spikes jutting from the back of its jaws and hair-like bristles, similar to the earliest form of feathers. Its skull was almost rounded and this, along with its beak, made it look similar to modern parrots (its name means 'parrot lizard'). *Psittacosaurus* is also famous for having more known species than any other type of non-bird dinosaur – so far 11 have been discovered in Asia. Hundreds of specimens have been found, many of them complete skeletons, and it is one of the most studied and known about dinosaurs.

2: Styracosaurus albertensis
Late Cretaceous, USA
Length: 5.5m; Weight: 3000kg
With five spikes sprouting from its neck frill, the tallest measuring 60cm, *Styracosaurus* looked ferocious. The spikes were purely decorative and most likely used to attract females.

3: Pentaceratops sternbergii
Late Cretaceous, USA
Length: 6.5m; Weight: 5000kg
Its name means 'five horned face' after the horns on its snout, eyebrows and cheeks. Its skull was up to 3m long, making it the longest skull of any land animal in history.

4: Triceratops horridus
Late Cretaceous, USA
Length: 9m; Weight: 11,000kg
As heavy as a truck, with an enormous neck frill and three impressive horns sprouting from its face, *Triceratops* was a powerful beast – and it needed to be. *Tyrannosaurus* was a known predator – its tooth marks have been found on *Triceratops* skeletons. It had one of the largest skulls of any land animal and unlike other ceratopsians, is thought to have lived alone rather than in herds.

Fighting Dinosaurs

In 1971, a Polish-Mongolian expedition found what is now one of the most famous specimens in palaeontology: two dinosaur skeletons locked together in a fight to the death, in the white sandstone cliffs of the Gobi Desert, in Mongolia. Evidence suggests these two dinosaurs were buried together, 74 million years ago, smothered by sand as a dune collapsed on top of them, either as a result of heavy rainfall destabilising the dune or from a sudden sandstorm. One was *Velociraptor mongoliensis*, a fierce predator; the other *Protoceratops andrewsi*, a small, sheep-sized ceratopsian.

In their lethal death grapple, *Velociraptor* has the sharp 'killing claw' of its left foot in the *Protoceratops'* neck, in the throat region, possibly striking a major artery, while its hindlimbs are kicking at the *Protoceratops'* chest and belly. In turn, the *Protoceratops* is biting down on the *Velociraptor*'s right arm to the point where it appears to be broken.

These fighting dinosaurs may have killed each other before being covered by sand, or died mid-fight as the sand suffocated them. Either way, this famous find captures a snapshot in time, revealing the ferocity and violence of the dinosaur world.

Key to plate

1: Velociraptor mongoliensis
Late Cretaceous, Mongolia
Length: 2.5m; Weight: 15kg
Part-scavenger, part-predator, *Velociraptor* would have mostly preyed on small animals, but was also an opportunistic feeder. In 2008, *Protoceratops* fossils were found with *Velociraptor* tooth marks scraped along the jawbone, as well as some *Velociraptor* teeth. The lack of meat on a jaw bone suggests the *Velociraptor* was gnawing on the *Protoceratops'* remains. In 2012, another *Velociraptor* fossil was found with a large pterosaur bone it its guts. With a wingspan of 2m, the pterosaur would have been too big

for a *Velociraptor* to kill, so it must have scavenged on the bones.

There is also good evidence to suggest that *Velociraptor* was nocturnal. Analysis of the bones around its eyes show that it could see and hunt well in the dark.

2: Protoceratops andrewsi
Late Cretaceous, Mongolia
Diameter: 1.8m; Weight: 180kg
A relative of the much larger *Triceratops*, *Protoceratops* had a frill on the back of its neck, but was without horns, although it did have two large bones jutting out of its cheeks. It is thought to have used its neck frill

either for courtship displays, for species recognition or to establish dominance in the herd.

Hundreds of *Protoceratops* fossils have been found together, so it seems likely that they lived in herds. Extensive fossil finds of *Protoceratops* at various life stages, from eggs to hatchlings to juveniles and adult males and females, have enabled scientists to build up a detailed picture of this dinosaur's life stages.

It is also known to have possessed strong jaws and a powerful bite, as seen in its death grip on the unfortunate *Velociraptor*.

Gallery 6

Non-Dinosaurs

Pterosaurs

The pterosaurs were winged reptiles, close cousins of the dinosaurs, and the first animals (after insects) to become capable of powered flight, flapping their wings to travel through the air. They were hugely successful, surviving from the Late Triassic to the end of the Cretaceous period and their fossils have been found on every continent. They evolved into a diverse range of species, from the size of a pigeon to that of a small plane.

Pterosaur wings were made up of a membrane of skin, muscle and tissue, which stretched from the ankle bones to an extremely elongated fourth finger. Many species were fast, agile fliers. When on land, most pterosaurs walked on all fours with their limbs tucked under their bodies rather than sprawled out to the side. Some were also efficient land predators, capable of walking, running and hunting on the ground.

Early pterosaurs were relatively small, with long tails, while later pterosaurs went on to develop elaborate head crests, specialized teeth or toothless beaks, and in some cases, enormous body sizes.

Compared to dinosaurs, their fossils are extremely rare, on account of their light, hollow bones, which were often too fragile to preserve well, and the fact that few lived in places where fossils easily form. However, the past decade has seen an exciting explosion of pterosaur finds, particularly in China and Brazil.

Key to plate

1: Quetzalcoatlus northropi
Late Cretaceous, USA
Wingspan: 12m; Weight: uncertain
One of the largest animals ever to fly. It had a huge toothless beak and a wingspan the length of a bus. The strength in its thickened arm bones enabled it to fly and it is thought to have taken off from a crouching position on all fours.

2: Caiuajara dobruskii
Late Cretaceous, Brazil
Wingspan: 2.4m; Weight: uncertain
A recently discovered species, found in the first ever pterosaur 'bone bed' made up of at least 47 individuals,
from juveniles to adults. They lived in colonies, flew at a young age and had a crest that changed from small and sloping in juvenlies to large and steep.

3: Eudimorphodon ranzii
Late Triassic, Italy
Wingspan: 1m; Weight: 10kg
Typical of early pterosaurs with a small wingspan, a short neck, sharp teeth and a long tail. It hunted for fish (scales have been found in its fossilised stomach area) and had more than 100 needle-like teeth.

4: Sordes pilosus
Late Jurassic, Kazakhstan
Wingspan: 0.6m; Weight: 5kg
The first pterosaur fossil to be found covered in hair-like fibres, with one specimen having a thick, fur-like coat. This covering was probably for insulation, suggesting that pterosaurs may have been warm-blooded.

5: Dimorphodon macronyx
Early Jurassic, England
Wingspan: 1.2m; Weight: 2kg
Its large clawed hands enabled this pterosaur to climb up steep cliffs so it could then launch itself off and fly. It had both fangs and grinding teeth, suggesting a diet of insects and small animals rather than fish.

Marine Reptiles

In the Mesozoic era, the seas and oceans were filled with a huge variety of large marine reptiles, including nothosaurs, ichthyosaurs, plesiosaurs and mosasaurs, all of which were only distantly related to dinosaurs.

The dolphin-like ichthyosaurs first appeared 245 million years ago. They dominated the seas during the Late Triassic and Jurassic before becoming extinct in the Late Cretaceous, some time before mosasaurs and plesiosaurs. They were superbly adapted to the water, with streamlined bodies and powerful tails and flippers. They gave birth to live young, so never had to leave the water to lay eggs, but lacked gills and needed to come to the surface to breathe. While most were around 3m in length, some reached the size of orcas. Early ichthyosaurs had long, flexible bodies, while later species were more compact and fish-like.

The other dominant group of marine reptiles were the plesiosaurs, which evolved from the nothosaurs at the end of the Triassic. Whereas nothosaurs had webbed feet, the plesiosaurs had two pairs of flippers. Many had extremely long necks, small heads and sharp, pointed teeth for catching fish. Another group, the pliosaurs, were the top underwater predators of their time, with huge heads and massive teeth. Their prey included large fish, ichthyosaurs, long-necked plesiosaurs and even dinosaurs scavenged from the shore.

Key to plate

1: Nothosaurus marchicus
Middle Triassic, Holland
Length: 1.5-2m; Weight: 80kg
The nothosaurs evolved in the seas at around the same time as the first dinosaurs appeared on land. They probably lived much like seals do today, feeding and breeding on rocks and beaches and diving into the water to feed on fish and shrimp.

2: Ichthyosaurus communis
Early Jurassic, England
Length: 2m; Weight: 90kg
A streamlined hunter, Ichthyosaurus was capable of catching fast, slippery prey in its needle-sharp teeth. Like other ichthyosaurs, it had large eyes protected by bony rings, which helped it to see in deep water.

3: Elasmosaurus platyurus
Late Cretaceous, USA
Length: 14m; Weight: 2000kg
One of the longest-necked creatures ever to have lived, this plesiosaur's neck was supported by 71 vertebrae. The neck itself was fairly rigid, but Elasmosaurus would have been able to twist and turn in the water in pursuit of its fishy prey.

4: Kronosaurus queenslandicus
Early Cretaceous, Australia
Length: 10m; Weight: 11,000kg
A pliosaur with a head about 3m long and a skull nearly twice the size of Tyrannosaurus'. It used its paddle-like flippers to 'fly' through the water. Fossil stomach contents have revealed that it fed on turtles and plesiosaurs.

5: Tylosaurus proriger
Late Cretaceous, USA
Length: 14m; Weight: uncertain
One of the last of the mosasaurs, Tylosaurus dominated the Late Cretaceous seas. Like other mosasaurs, it swam like a crocodile, moving its body in slow waves, although it was capable of sudden bursts of speed. It had double rows of pointed teeth lining its jaws and was a deadly predator.

Mesozoic Mammals

The ancestors of modern mammals came from a group of mammal-like reptiles known as the cynodonts, which first appeared 260 million years ago and survived into the Triassic. They walked on four legs and may have been covered in fur. The word 'cynodont' means 'dog teeth' as, like their mammal descendants, they had differentiated teeth. They were also warm-blooded and had large brains for their body size.

The first mammals to evolve from the cynodonts were small, shrew-like creatures that first appeared in the Triassic period. Many of them laid eggs, like their reptilian ancestors, and hunted at night, feasting mainly on insects. But as the Mesozoic era progressed these early mammals began to diversify, eating both meat and plants, and living in water as well as on the ground and in trees. *Castorocauda*, for example, a protomammal from the Middle Jurassic, had a beaver-like tail, limbs adapted for swimming, and teeth well-suited to eating fish. *Volaticotherium*, from the Jurassic, is the earliest-known gliding mammal with a membrane stretched between its limbs, similar to a modern flying squirrel. Recent finds from China have shown that some mammals from the Cretaceous period were much larger than previously thought, reaching lengths of 1m and more, and even preyed on baby dinosaurs.

Key to plate

1: **Repenomamus giganticus**
Early Cretaceous, China
Length: 1m; Weight: 14kg
A raccoon-like, muscular animal, far heftier than other mammals known from this time. One was found with a fossilised young *Psittacosaurus* in its gut, although it's not known if it actively hunted dinosaurs or scavenged them.

2: **Cynognathus crateronotus**
Early–Mid–Triassic, Africa, South America and Antarctica
Length: 1.2m; Weight: 6.5kg
A fast, fierce predator. This cynodont was heavily-built with wide jaws and sharp teeth capable of slicing through

flesh. It was possibly warm-blooded and covered in hair.

3: **Megazostrodon**
Late Triassic–Early Jurassic, South Africa
Length: 10cm; Weight: 28g
A tiny, shrew-like creature with most of its length made up by its tail. It probably lived in underground tunnels and may have eaten roots or burrowing insects. It had a well-developed sense of smell and hearing.

4: **Eomaia scansoria**
Early Cretaceous, China
Length: 14cm; Weight: 20-25g
Its name means 'dawn mother' as

it is the earliest-known ancestor of placental mammals. It was tiny, hairy, well-adapted for scrambling about in bushes and trees and probably fed on insects.

5: **Volaticotherium antiquum**
Mid–Late Jurassic, China
Length: 30.5cm; Weight: uncertain
With its grasping toes and membrane between its limbs, *Volaticotherium* would have been able to climb trees and then glide between the branches. It was also fur-covered, with specialised teeth for feeding on insects.

Extinction

Around 66 million years ago, non-bird dinosaurs became extinct, along with more than half the world's animal species, including mosasaurs, plesiosaurs, sea sponges and ammonites, and much of the world's vegetation.

Some evidence shows that the dinosaurs might have been in decline long before the extinction event, while other evidence suggests that specialist groups were flourishing in some parts of the world. So it would seem the dinosaurs were not doomed to extinction. Whatever wiped them out was sudden and catastrophic.

There are two main theories as to what happened. Today, most scientists believe the dinosaurs were wiped out by a 9.7km-wide asteroid, travelling 20 times faster than a speeding bullet, which collided with the Earth in what is now Mexico's Yucatán Peninsula. In the immediate aftermath of the impact, tsunamis would have hit the world's coasts, while wildfires scorched the land. Clouds of dust would have blocked out sunlight, causing temperatures to plummet around the globe. Plants would have been unable to photosynthesise, killing herbivores and, in turn, starving carnivores. Evidence for the asteroid theory comes from the 180km-wide Chicxulub crater, and layers of rock from around the world rich in the metal iridium, which is rare on Earth but is commonly found in meteorites. Both the crater and the layers of iridium date back to the exact time of the extinction event.

Other scientists lay the blame on a bout of volcanism in India's Deccan Traps. Prolonged and massive volcanic activity could also have spread iridium around the world, as iridium is rich in the Earth's core. Exploding volcanoes would have blocked out sunlight as they spewed ash and dust into the atmosphere, as well as causing climate change from greenhouse gases. Or, others have argued, both volcanism and the asteroid impact could have worked together – a deadly cocktail to wipe out the dinosaurs' reign on Earth.

Key to plate

1: **After the impact**
This image depicts the world as it may have looked in the aftermath of an asteroid collision with Earth 66 million years ago. Clouds of dust fill the sky, blocking out sunlight. The ground is laid bare, much of Earth's plant life having withered and died, and with it, many of the animals that relied on vegetation to survive. It would have been a time of cold and darkness – a winter on an epic scale.

Survivors

A third of all living creatures survived the catastrophe that wiped out non-bird dinosaurs. Small lizards, snakes, birds, insects, mammals, sharks, turtles, amphibians and crocodiles were among those that made it through what is now known as the Cretaceous-Palaeogene (K-Pg) extinction event. But why did some animals survive and not others?

Crocodiles would have been able to survive thanks to their ability to go long periods without eating and to migrate to better living conditions. Birds were able to fly around the globe in search of food, while smaller animals would also have been at an advantage. They could have burrowed to survive the harsher conditions and were also more likely to have had a varied diet, rather than relying on one food source, and would have only needed small quantities of food each day.

For the survivors, the mass extinction event was also an opportunity. All major extinctions of life on Earth have been followed by a burst of evolution, as animals diversify to occupy the habitats left behind by their predecessors. And while birds spread around the world, it was mammals that really gained from the dinosaurs' demise.

In the 20 million years that followed the K-Pg extinction, mammals greatly diversified and sometimes vastly increased in size. No longer small and nocturnal, hiding in the undergrowth, they slowly began to dominate almost every habitat, with some even taking to the oceans. By the end of the Palaeogene period that followed, 23 million years later, mammals had diversified into primates, horses, bats, pigs, cats, dogs and whales. The 'Age of Mammals' had begun.

Key to plate

1: **Hyracotherium**
Eocene, North America and Europe
Length: 78cm; Weight: 9kg
Known as the 'dawn horse', *Hyracotherium* is the ancestor of modern horses. It lived around 50 million years ago, a browsing herbivore feeding on soft plants.

2: **Moeritherium**
Eocene, Africa
Height: 70cm; Weight: 235kg
A pig-like mammal related to

elephants that lived about 35 million years ago, in swamps and rivers.

3: **Gastornis**
Late Palaeocene and Eocene, Europe, China and possibly USA
Height: 2m; Weight: 170kg
A large, flightless bird with a powerful beak. It may have been an ambush hunter or fed on large vegetation.

4: **Palaeotrionyx**
Palaeocene, North America

Length: 45cm; Weight: 6kg
This soft-shelled freshwater turtle would have looked very similar to its modern relatives, with a long neck, sharp beak and three-toed feet.

5: **Dorudon**
Palaeocene, seashores of North America, northern Africa and the Pacific Ocean
Length: 5m; Weight: 450kg
An ancient toothed whale that fed on fish and molluscs.

DINOSAURIUM

Library

Index

Curators

Chris Wormell is a self-taught artist and celebrated printmaker. He creates his timeless illustrations using two main methods: wood engraving and linocut. Chris has written and illustrated many children's books and most recently gained recognition for the cover illustration of the award-winning bestselling *H is for Hawk* as well as for Philip Pullman's new book *La Belle Sauvage: The Book of Dust Volume One*.

Lily Murray has worked as an editor and writer for over 15 years. She enjoys walking, birdwatching, trips to the Natural History Museum and reading about new dinosaur discoveries. She is also a (very amateur) fossil hunter.

Dr Jonathan Tennant completed his PhD at Imperial College London in 2016, where he researched the evolution and extinction of dinosaurs. He has written, and been consultant on, several books and is a freelance science writer.

To Learn More

University of California Museum of Paleontology
A fantastic site with introductory information and facts about dinosaurs.
www.ucmp.berkeley.edu/diapsids/dinosaur.html

The Smithsonian Museum of Natural History
Information about the museum's dinosaur fossil collections. Click through to their latest exhibitions with featured videos and blogs.
www.naturalhistory.si.edu

The Melbourne Museum, Australia
Take a Dinosaur Walk and find out more about the megafauna of the Mesozoic.
www.museumvictoria.com.au/melbournemuseum/discoverycentre/dinosaur-walk

BBC Nature: Prehistoric life
See fantastic dinosaur reconstructions, learn new dinosaur facts and watch video clips on the BBC website.
www.bbc.co.uk/nature/life/Dinosaur

The Natural History Museum, London
An A–Z guide to dinosaurs, with images, facts and figures for over 300 dinosaurs.
www.nhm.ac.uk/discover/dino-directory/index.html

Paleobiology Database
A massive database of all known dinosaur species, including an interactive map to explore fossil sites around the world.
www.paleobiodb.org/navigator

American Museum of Natural History
See online exhibitions from one of the greatest dinosaur fossil collections in the world and watch videos of museum scientists answering questions about dinosaurs.
www.amnh.org/dinosaurs

Australian Museum
Find out about Australian dinosaurs and the evolution of them into birds.
www.australianmuseum.net.au/dinosaurs-and-their-relatives

Dinosaur News
Catch up on all the latest dinosaur developments.
www.dinosaurnews.org

I KNOW DINO! The big dinosaur podcast
Podcasts with news, interviews and discussions about dinosaurs.
www.iknowdino.com

Discovery kids: Dinosaurs
Dinosaur website for children with quizzes, games and videos.
www.discoverykids.com/category/dinosaurs

Dinosaur Fact
One of the largest compilations of dinosaur facts on the web. You can also click through to see a complete dinosaur cladogram.
www.dinosaurfact.net